How to Be Human

For the past 60 years ***New Scientist*** has been reporting on the extraordinary ingenuity of nature and of humans. Its accounts of the latest discoveries and inventions, why they matter and their implications, has made the magazine the world's bestselling science weekly. And through its books, apps, live events and website it reaches more than 5 million people worldwide.

Graham Lawton

After a degree in biochemistry and a MSc in science communication, both from Imperial College, Graham Lawton landed at *New Scientist*, where he has been for almost all of the twenty-first century, first as features editor and now as executive editor. He wrote and edited the previous *New Scientist* illustrated book *The Origin of (almost) Everything*. His writing and editing have won a number of awards.

Jeremy Webb

In his varied career Jeremy Webb has been a sound engineer, radio producer and TV news editor. At *New Scientist* he's been Editor, Editor-in-Chief and now Editor-at-Large. For a decade he co-edited 'The Last Word' column, which spawned bestselling books such as *Why Don't Penguins Feet Freeze?* and *Nothing*. He has a degree in physics from the University of Exeter.

Jennifer Daniel

Jennifer Daniel is an award-winning illustrator and author who regularly contributes to the *New York Times* and the *New Yorker*, and creates animations and illustrations for various publishers. She is a Creative Director at Google, former graphics editor at the *New York Times* and has been recognised with many fancy awards for her work in visual styling.

How to Be Human

NewScientist

Words by
Graham Lawton & Jeremy Webb

Illustrations by
Jennifer Daniel

Contents

Introduction

Take a look at yourself

A few years ago – more than either of us would probably care to admit – we both attended a *New Scientist* event on the nature of animal minds. In one memorable talk, an eminent biologist described how sheep get nervous if they are left alone – they evolved to live in flocks, after all. His remedy? Put a mirror in the pen. It fools them into thinking they have company.

Sheep are not much blessed with self-awareness. Like most other species on Earth, they cannot recognise themselves in a mirror. But, as a member of the planet's smartest species, you are different. You know it's you gazing back out of the glass.

This rare ability is shared by only a handful of brainy species. In humans it's a milestone of infant development, showing that a child is acquiring a sense of self. In turn, this helps to kick into action memory and important social skills, such as reading the intentions of others.

Your own memory, sense of self and mind-reading abilities have been crucial in shaping who you are. And 'you' is what this book is all about – how you got to be who you are and what makes you tick. We'll also look more widely at 'us': what sets humans apart from other animals and how we can be so similar yet individually unique.

The nature of humans has intrigued scholars for many thousands of years. The ancient Greek philosopher Socrates is said to have started the ball rolling, though it was his successors Plato and – especially – Aristotle who had the most influence on subsequent thought. But Aristotle, as usual, got plenty of things wrong: he believed that the heart was the seat of thought and reason and the brain was for cooling the blood, for example. That view held for nearly 2,000 years until William Harvey discovered that the heart is simply a mechanical pump.

As with many discoveries in science, that brought our view of ourselves down a peg or two. We once considered ourselves to be a special (though fallen) creation of God, made in His image and living at the centre of a cosmos made for us; nowadays, we're more likely to understand ourselves as an accidental product of evolution living in an uncaring cosmos with only our deeply flawed brains to guide us.

But at least we have retained our sense of humour. As the American essayist Christopher Morley once said, a human being is basically 'an ingenious assembly of portable plumbing'. There's more to us than that, of course, and the discoveries of science have uncovered a creature way more fascinating than any theological ideal. Who would have thought, for example, that laughter has less to do with humour than it has social control?

So what is a typical human like?

There's no doubt we're smart. Our big brains are the main reason we live such different lives from the rest of Earth's inhabitants; they have enabled us to go from savannah-dwelling hunter–gatherers to the creators of a civilisation with aspirations to cure death, colonise other planets and build machines that are even smarter than us.

We've been on a 7-million-year evolutionary journey from when we last shared an ancestor with chimpanzees. Darwin's big idea has proved crucial to understanding ourselves, from why each of us has a different personality, where our morals come from, to why only some of us can digest milk.

Being upright, and not walking on all fours, enabled our hands to become strong, dexterous, multifunctional tools perfect for turning thoughts into things, such as stone axes, music, skyscrapers and text messages. It also allowed us to evolve into supreme endurance predators, able to run quarry to exhaustion. Without this bounty of high-quality protein it is unlikely that our ancestors would have evolved their big brains.

One trait we tend to overlook is our predilection for possessions, yet it is a defining characteristic of our species. Early humans would probably not have migrated from Africa without warm clothing, tools for hunting, gathering and preparing food and the means to make fire. No other animal depends on so much stuff to survive. It is also via possessions that we express many other facets of our character including status, symbolism and aesthetics.

We're also incorrigibly talkative. We are the only creatures with complex language, which has enabled us to create knowledge, add to it and pass it on to others. Language also facilitates our intensively social nature: to be talkative we need somebody to talk to. We thrive in the presence of friends and family and wither without them. But language also helps to divide us into mutually unintelligible and often hostile tribes.

If this description sets the scene for this book, the drama itself focuses on the human condition, the experience of being alive and the events and stages we face in our journey through life. You'll find here what science tells us about generosity, belief, disgust, why we pick up bad habits and find it hard to kick them. We'll also cover life's phases from birth to death, how children change their parents and the upsides of old age.

Understanding humans and their place in the cosmos is fundamental to what we do at *New Scientist*. The text of *How to Be Human* owes so much to the insatiable curiosity and expert knowledge of our brilliant colleagues, and to the dozen guest authors, who write here on topics such as the nature of friendship and why we get hooked on religions. When it comes to illustrations, Jennifer Daniel has provided highly entertaining infographics covering everything from the curious concept of personal space to the link between Stephen Hawking and Black Sabbath; while Kirstin Kidd has found fascinating photographs revealing what it means to be human that range from 30,000-year-old art right up to the latest technology for freezing people who want a second chance at life. We only hope you enjoy reading this book as much as we've enjoyed writing it.

This is undoubtedly a great time to be writing about humans. Science is generating so many insights on so many fronts. Take for example, the sequencing of the human genome which is not only revolutionising medicine and human biology but it's also revealing home truths about our ancestors, who clearly enjoyed sexual liaisons with Neanderthals.

And in a few decades, the human mind has been transformed from an inscrutable black box into a series of spaces wide open for interrogation. Neuroscientists and psychologists can now see thoughts form in the brain and reliably predict the mistakes people will make from biases deep in their unconscious.

Let's finish with a confession. Our choice of book title *How to Be Human* might imply that we have (or think we have) all the answers. We don't. This is not an instruction manual or self-help book. Rather, it describes how far science has come in answering some of our biggest questions about ourselves, and many of the small ones too. As Socrates himself said, an unexamined life is not worth living. So get to know yourself better.

Graham Lawton and Jeremy Webb
June 2017

1
Human Nature

What sort of creatures are we?

If alien biologists visited Earth they would certainly notice us, and would have no trouble identifying our defining characteristics. Their list would include intelligence, language, sociability, religiosity, technological prowess and a liking for material possessions. These characteristics are at the core of what we call human nature.

But they are far from the only ones. Humans typically display many other, less obvious, traits that maybe we are too close to the subject to see. If you thought you knew what humans were like, then think again.

The first of the less obvious traits that our alien biologists would note is playfulness. Even before we walk or talk, we play. It comes naturally to us and occupies much of our childhood.

Playfulness itself doesn't make us unique: all mammals, some birds and a few other animals play. But no other species pursues it with such variety and vigour. The ways we play include such activities as games, jokes, sports, music, dancing, art and plain old horsing around. We play with each other, with objects, with language and with our imaginations. We play in the real world and in virtual ones. We are also unusual in that we retain our juvenile sense of fun into adulthood. The only other primate to do that is the bonobo.

Why do we do it? One factor may simply be the availability of leisure time. In the wild, adult chimps spend around 8 hours a day foraging. Given more free time, they play more. However, play isn't simply a way of whiling away the hours. It is also an evolutionary adaptation for learning. The adult world is socially and physically complex, and play is largely a preparation for entering it.

Purposeful play

Four primary purposes have been identified for play – physical development, social development, hand–eye coordination and training for the unexpected. Not for nothing is it called 'the work of the child'.

Very superstitious

Barack Obama used to play basketball on the morning of an election. Golfer Tiger Woods wears a red shirt when competing on a Sunday. Most of us have superstitions, even though we know rationally that they cannot work. Yet superstition is not entirely nonsensical. Our brains are designed to detect patterns and order in our environment and to assume that outcomes are caused by preceding events. Both abilities evolved for good reason. Our ancestors would not have lasted long if they had assumed that a rustling bush was caused by the wind rather than a lion. But this survival adaptation leaves us wide open to misattributing effects to causes, such as a football team winning because they're wearing lucky underpants. In other words, superstition.

As well as playing, young humans also spontaneously embark on another quest that marks us out from other animals. From earliest infancy, we constantly seek to sort the world into categories, try to understand how things work, make predictions and test them. This is the quest for knowledge, and it is evident in a range of human inventions from time, calendars and cosmology to family names and measurement. It is also the essence of science.

Abstract thinking

Again, it is not totally unique. All animals need thinking of this sort to survive. Pigeons, for example, can learn to discriminate between cars, cats and chairs. Dogs can associate the sound of a bell with food, and when chimps try to extract a nut from a tube, they are performing a simple experiment.

Clearly, though, no other animal does science to the extent that we do. What sets us apart is the ability to grasp abstract concepts. Chimps struggle with this. For example, while they quickly learn that heavy rocks are better for smashing nuts, when it comes to a general understanding, they fail. If they hear two objects drop and one goes 'crash' while the other goes 'clack', they can't infer that one of those objects will be good for cracking a nut and the other won't. Crucially, this understanding allows us to use what we have learned in one domain to make causal predictions in another. We can predict that the object that is good for crushing nuts will probably sink and the other float, for example.

Another trait that distinguishes us from less scientific animals is an eagerness to share what we have discovered. Once we figure something out, we announce it to the world, which is why all scientifically minded humans, not just Newton, can stand on the shoulders of giants.

Aside from discovering the laws of nature, we also have an urge to create laws to regulate human interaction. This 'legislativeness' is another thing that defines us as a species.

The question of whether every human society has formal laws is far from settled, but they do all have rules. This is a peculiarly human trait. Chimps stick to simple behavioural rules governing things like territory and dominance hierarchies, but we have developed much more elaborate systems of rules, taboos and etiquette to codify behaviour. Though every society has different rules, they always involve regulating activity in three key areas – a sure sign that these are fundamental to human nature.

First and foremost are rules about kinship, including who counts as kin and the rights and obligations it confers in relation to the inheritance of goods and status. Every society also recognises the concept of kinship by marriage, as well as incest taboos prohibiting sexual intercourse between immediate family (though royalty is sometimes exempted).

Drawing the line

After who's who, the next concern is safety, so every culture also has rules about the circumstances under which one person can kill another. Condemnation of murder is universal, though its definition varies. In some societies, any stranger is fair game. Others allow killing to avenge the murder of kin, and many allow the group to kill someone who violates its norms. But every group draws the line somewhere.

Every society also has rules governing access to material goods. The notion of private property is by no means universal, but people everywhere have rules that stipulate who is entitled to use certain things at certain times, as well as punishments for people who break them. These vary widely from a simple first come, first served, to elaborate systems of private ownership.

Secret sex

Kinship, safety, stuff. Across the whole range of human cultures this is what our rules say we care about. Another is sex. Nothing reveals an animal's nature quite as well as its sexual practices, and humans certainly have some odd ones. Women are continually receptive and have concealed ovulation – that is, there is no external sign that they are ready to conceive. We are generally monogamous but live in large mixed-sex groups, which is unique among primates. But surely nothing is quite as puzzling as our predilection for clandestine copulation. Why do humans have sex in private?

This coyness is not just a product of specific cultures or morals; it is the rule across all societies. There is the odd instance of public ritual sex, and drunken orgies are certainly not unknown. But where there is no alcohol – as would have been the case in the past before agriculture – sexual privacy is the norm.

To see how unusual that is, consider our close relatives. Among orang-utans and gorillas, where there is a lot of inter-male competition, alpha males copulate openly. In bonobos, sex is a free-for-all with no privacy sought whatsoever.

Our innate demand for privacy probably evolved in response to our increasingly complex sexual politics. For a start, women won some control from men by evolving concealed ovulation and continual sexual receptivity to confuse paternity. Then our ancestors did something completely different from other great apes – males and females started sharing parental care. Monogamy was born, and along with it the need to strengthen the pair bond. Privacy may have emerged as a way to increase intimacy.

But as well as strengthening relationships, clandestine mating also makes it easier to get away with infidelity. And infidelity appears to be very common. It is widespread in all traditional cultures, and private sex allows it to occur without loss of reputation.

Another very human trait, envy, may also play a part. Since men can never get enough of it, sex is a precious commodity and therefore best enjoyed covertly to avoid inciting covetousness. Like food in a famine, somebody who has plenty would be wise to eat it in private. A sexual act, even among consenting adults, has a high probability of upsetting someone. Parents or community members may disapprove and for children it can lead to the creation of rival siblings. So perhaps clandestine copulation simply follows the precautionary principle.

Making a meal of things

Food is another of life's pleasures, and compared with other animals, the feeding behaviour of humans is odd. Where they just eat, we make a meal of it. The main difference is down to one of humanity's greatest inventions: cooking. People in every culture cook at least some of their food.

Culinary culture includes the strange phenomenon of ritualised, familial food-sharing, otherwise known as mealtimes. Chimps eat their food individually as they find it throughout the day. But in every human society, people gather in family groups at more or less regular times of day to eat food that has been prepared communally, usually by women. There's also feasting. From sharing the spoils of a good hunt to celebrating a special occasion, every society does it. And here you are more likely to find men cooking. We even see this in our own backyards, where they do most of the barbecuing. It may have something to do with establishing status, by being generous with high-quality food.

For humans, eating is about much more than nourishment. Food is used to form social bonds. Mealtimes are the centrepiece of family life; feasting bonds friends, colleagues and communities; and we also use food to consolidate more intimate relationships.

Gossip's hidden value

Mealtimes are also full of something else that defines our species: chatter. Language was once thought to be the defining characteristic of humans. These days we are more likely to consider it as part of a continuum of animal communication. Nevertheless, nobody doubts that it has shaped our nature profoundly. Language is central to many human universals ranging from education, folklore and prophecy to medicine, trade and insults. Arguably, though, our way with words reaches its apogee in something that is often considered trivial or even shameful: gossip.

A compulsion to talk about other people is only human. And it is not nearly as frivolous as you might think. Some anthropologists believe we gossip to manipulate the behaviour of others, which may help to explain why gossip often takes place within earshot of the person being gossiped about. Among the Kung Bushmen of Africa, for example, that is the case 70 per cent of the time.

But gossip doesn't just serve to name and shame. Barbed comments are relatively rare compared with innocuous ones. Gossip may be the human equivalent of primate grooming – our social relationships are too numerous to cement each one with time-consuming grooming, so we chat instead. Even the most powerful movers and shakers depend on it, though they may call it by some other name. After all, most business could easily be transacted by phone or email, but people still meet face-to-face so that they can bond over casual conversation at lunch or on the golf course.

A juicy titbit of gossip is like a gift – and, incidentally, gift-giving is another human universal. Both are like glue that binds societies together. A society without gossip may simply dissolve, as people wouldn't have any common interest in staying together.

Arty farty

Explaining the human urge to create works of art is a challenge. Darwin suggested it has its origins in sexual selection: like a peacock's tail, creativity is a costly display of evolutionary fitness. When women are at their monthly peak in fertility, they prefer creative men over wealthy ones. But sex alone may not explain the evolution of art. Another idea is that the drive to seek out aesthetic experiences evolved to push us to learn about different aspects of the world. Art is a form of intellectual play, allowing us to explore new horizons in a safe environment.

Why do we laugh, blush and kiss?

Human behaviour is extremely flexible and varied. Most of what we do serves an obvious function: we eat, sleep, talk, groom, have sex, travel, work, exercise, entertain ourselves and, given half a chance, loaf around. But there are still some corners of our behavioural world that we really do not understand.

One of our most mysterious acts is blushing. In a species with a reputation for cunningly manipulating others to maximise personal gain, reddening of the cheeks is difficult to explain. Why would humans evolve a response that puts us at a social disadvantage by forcing us to reveal that we have cheated or lied?

It is a question that Charles Darwin struggled with. He pointed out that while all people of all races blush, animals – other primates included – do not. When it came to explaining the evolution of 'the most peculiar and the most human of all the expressions', he was at a loss. That has not stopped others from trying.

A show of honesty

One suggestion is that blushing started out as a simple appeasement ritual: a way to show dominant members of the group that we submit to their authority. Perhaps later, as our social interactions became increasingly complex, it became associated with higher, self-conscious emotions such as guilt, shame and embarrassment. This would seem to put individuals at a disadvantage, but blushing might actually make a person more attractive or socially desirable.

That may be because it is a hard-to-fake signal of honesty. Women blush more than men, leading to the suggestion that blushing evolved as a way for women to demonstrate their fidelity to men and so enlist their help in rearing offspring. Blushing says 'I can't cuckold you. If you ask me about infidelity, I can't lie – my blush would give me away.'

Similarly, blushing could have emerged as a way to foster trust. Once blushing became associated with embarrassment, anyone who did not blush might have been at a disadvantage because we are less likely to trust someone who appears never to feel ashamed about anything.

Embarrassing situations are also likely to elicit another peculiar human behaviour: laughter. You may think it is obvious why we laugh. But most laughter has nothing to do with humour.

Who needs humour?

Starting in the late 1980s, psychologist Robert Provine of the University of Maryland carried out a 10-year study of laughter in various natural habitats. He visited shopping malls, school classrooms, sidewalks, offices and parties. After recording more than 2,000 instances of laughter, he concluded that laughter is prompted less by amusing jokes than by banal comments. Statements such as 'Do you have a rubber band?', for example, which was enough to make someone in a Baltimore shopping mall chuckle.

Provine's conclusion was that the essential ingredient for laughter is not humour but other people. Laughter is a social signal that we use in all sorts of situations to bind ourselves together.

Conversational laughter acts as a social lubricant. It engages listeners and dispels tension, aggression and competition by putting people at ease. Nervous laughter can make light of a stressful or psychologically difficult situation. And, through its contagious nature, laughter can unify the mood and behaviour of a group, promoting coordinated activity for the greater good.

Pick that one out!

In 2001, two researchers at the National Institute of Mental Health and Neurosciences in India won an Ig Nobel prize for their research into a very odd behaviour. Chittaranjan Andrade and B.S. Srihari asked 200 adolescents about their nose-picking habits and found that almost all of them admitted doing it an average of four times a day. Nine pupils even owned up to eating their bogeys. Why would anyone do that? Andrade points out that there isn't any nutritional content in mucus. It is possible that ingesting nasal detritus might help build a healthy immune response: lack of exposure to infectious agents is known to increase susceptibility to allergies. The only actual research on nose-picking dates back to 1966, when Sidney Tarachow of the State University of New York found that people who ate their bogeys found them 'tasty'. Yum . . .

Laughter can also be used as a tool of social control. As we master its subtle cues we begin to use it to manipulate those around us. An 'in' joke can exclude outsiders from a clique. Laughter can be used to show who is boss and malicious laughter is an effective weapon of intimidation. Laughing at someone rather than with them can pressure them to conform or push them away.

Another strange behaviour that usually needs more than one person to be present is kissing. It is not practised in all cultures, so it cannot be in our genes. So why do so many of us lock lips, and why does it feel so good? There is no shortage of speculation.

One idea is that our first experience of comfort, security and love comes from the mouth sensations associated with breastfeeding. In addition to this, our ancestors probably weaned their babies by mouth-to-mouth feeding of chewed food, as chimpanzees and some mothers do today, reinforcing the connection that people feel between sharing spit and pleasure.

Another idea is that kissing has its origins in foraging. Our ancestors were first attracted to ripe, red fruit, then co-opted this attraction for sexual purposes, developing pronounced red colouration on genitals and lips. When it comes to the physiology of kissing, we are on slightly firmer ground. Our lips are among the most sensitive body parts, packed with sensory neurons linked to the brain's pleasure centres. Kissing has been shown to reduce levels of the stress hormone cortisol and increase the bonding hormone oxytocin.

There may even be a link between kissing and the way we assess the biological compatibility of potential partners. Research conducted in recent years has revealed that we are most attracted to the smell of sweat from people whose immune system is most dissimilar from our own – with whom we are likely to produce the healthiest children. And of course kissing allows us to get up close and personal enough to sniff that out.

What has language ever done for you?

Try striking up a conversation with a newborn baby and you'll find it gets very one-sided very quickly. But try again three or four years later and you may struggle to get a word in edgeways. It won't be the most sophisticated conversation, but it will be better than you'll ever get out of a cat, dog, or even a talkative parrot. An average four-year-old has a vocabulary of about 5,000 words and sophisticated knowledge of how to string them together. Incredibly, nobody teaches them how to do it.

Our facility for language is perhaps the defining feature of our species. Wherever you find people you also find language, both spoken and written. We acquire it effortlessly as toddlers and use it every day of our lives. Without it, trade, tribes, religions and nations couldn't have existed, and civilisation as we know it would be impossible.

The language instinct is almost irrepressible. When people with no shared language are thrown together, they quickly invent a rudimentary system of verbal communication called a pidgin. Within two to three generations these can evolve into fully fledged languages. Deaf people have also spontaneously invented new sign languages.

Linguists define language as any system that allows the free and unfettered expression of thoughts into signals, and the conversion of such signals back into thoughts. This sets human language apart from all other animal communication systems. A dog's bark, for example, can only convey a limited amount of information: that it is hungry or has spotted an intruder. But cannot tell the story of its puppyhood or describe the route of its daily walk. Human language is unique in that it can convey almost any idea or event, even impossible ones.

Today, 7,000 or so languages are spoken – including sign languages – and countless more have gone extinct. Despite their obvious differences, deep down they are all the same, in that they can communicate the full range of human experience. That suggests language evolved long before our ancestors migrated out of Africa and around the globe starting some 100,000 years ago.

Social glue

Language was clearly one of the adaptations that allowed us to conquer the world and beyond. It is intimately tied to our intensely social and usually cooperative nature and probably evolved in parallel with it. Human societies are glued together by language. Traditional hunter–gatherer groups typically live in bands of about 100–150 people, which is too many to maintain good relations through actual contact. So we talk about one another instead. Language also allows us to exchange favours, goods and services, often with others outside our immediate family. These complicated social acts require more than grunts.

If language defines us as a species, it also moulds us as individuals. Back in 1940, linguist Benjamin Lee Whorf proposed that the language we speak influences how we see the world. He suggested, for example, that people whose languages lack words for a concept would not understand that concept. The idea was relegated to the fringes until the early 2000s, when a few researchers began probing a related but more nuanced idea: that language can influence perception.

Greek, for example, has two words for blue – *ghalazio* for light blue and *ble* for dark blue. Greek speakers can discriminate shades of blue faster and more accurately than native English speakers.

Language also affects our sense of space and time. For English speakers, time flows from back to front: we 'cast our minds back' and 'hope for good times ahead'. The direction in which our first language is written can also influence our sense of time, with speakers of Mandarin more likely to think of time running from top to bottom than English speakers. Some peoples, like the Guugu Yimithirr in Australia, don't have words for relative space, like left and right, but do have terms for north, south, east and west. They tend to be unusually skilled at keeping track of where they are in unfamiliar places.

The language we speak may even affect who we are. Neuroscientists and psychologists are coming to accept that language is deeply entwined with thought and reasoning, leading some to wonder whether people act differently depending on their language. Research in bilinguals suggests they do. In the 1960s sociolinguist Susan Ervin-Tripp asked Japanese–English bilinguals to complete unfinished sentences in both languages. Her subjects used very different endings depending on the language they were using. Given the sentence 'Real friends should ...' a typical Japanese ending was '... help each other out,' but in English it was '... be very frank'. Overall, the responses reflected how monolinguals of either language tended to complete the task.

Differences in attitude

Another experiment had bilingual English–Spanish volunteers watch TV adverts, first in one language, then in the other, and then rate the characters. When the ads were in Spanish, volunteers tended to rate the women as independent and extrovert, but after seeing them in English they described the same characters as hopeless and dependent.

Similarly, bilingual Mexicans describe their own personalities differently in different languages. In Spanish they tend to be more humble than in English, perhaps reflecting divergent US/Mexican attitudes to assertiveness.

Language can even shape your memory. Spanish speakers are worse at remembering who caused an accident than English speakers, perhaps because they tend to use passive phrases like '*Se rompióel florero*' ('the vase broke') that do not specify the person behind the event. To a large extent, you are what you speak.

Why are there so many languages?

If language evolved for communication, how come most people cannot understand each other? The Old Testament story of Babel explains it as God's way of preventing humans from becoming too powerful, by stopping them from cooperating. Oddly, this may not be far from the truth.

By nature, we are tribal and xenophobic, affiliating with our in-group and fearful and disdainful of outsiders. Language is both a powerful marker of tribal identity and a cypher that can prevent outsiders from eavesdropping. There are many recorded instances of groups deliberately changing their language to include insiders and exclude outsiders, and this may be a powerful driver of linguistic divergence.

What are you like?

When it comes to personalities, no two people are alike. Yet psychologists reckon they can describe the complete spectrum of individuality with just five broad traits. Each of the 'Big Five' - openness to experience, conscientiousness, extroversion, agreeableness and neuroticism - defines an axis along which we all fall

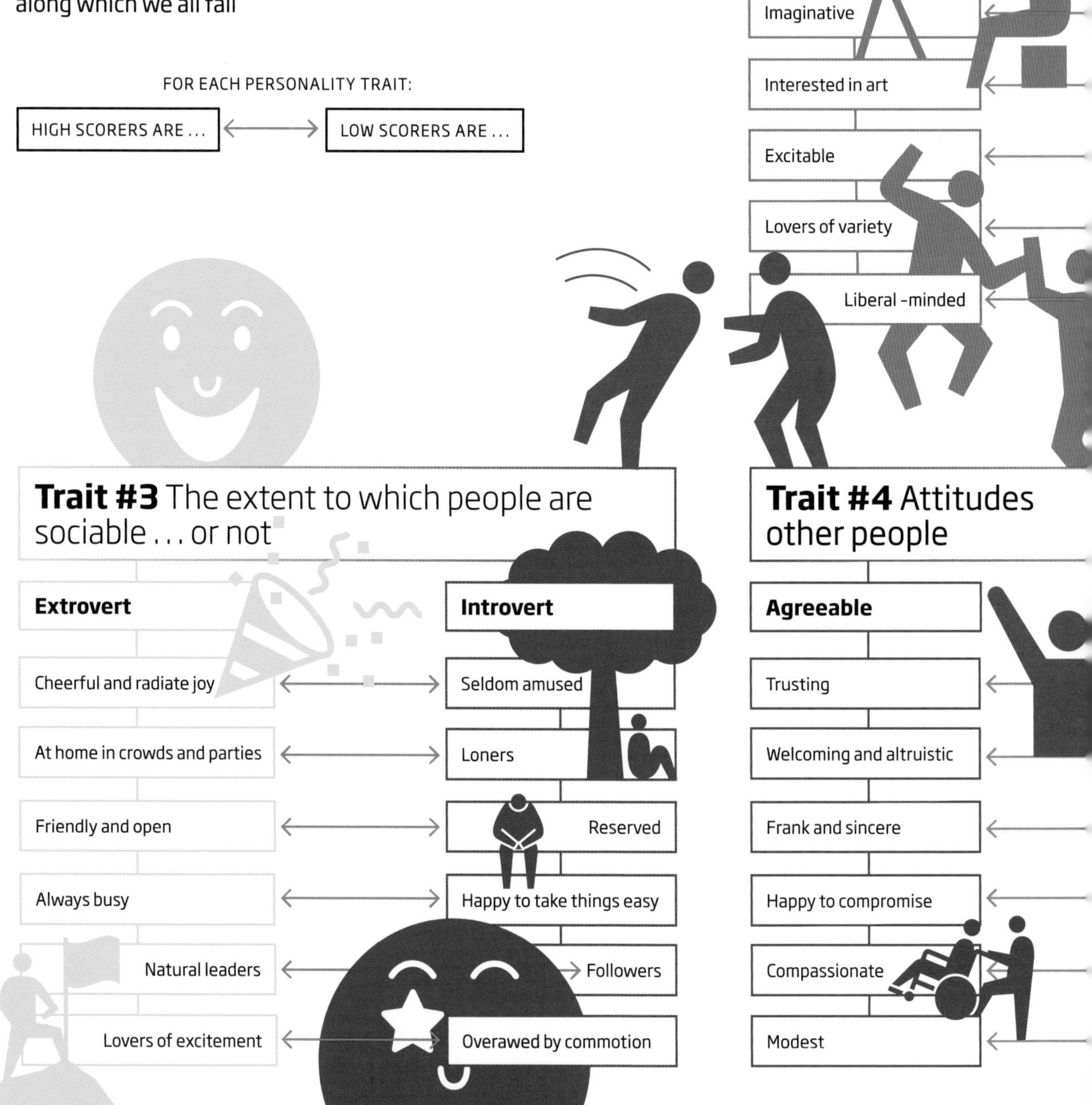

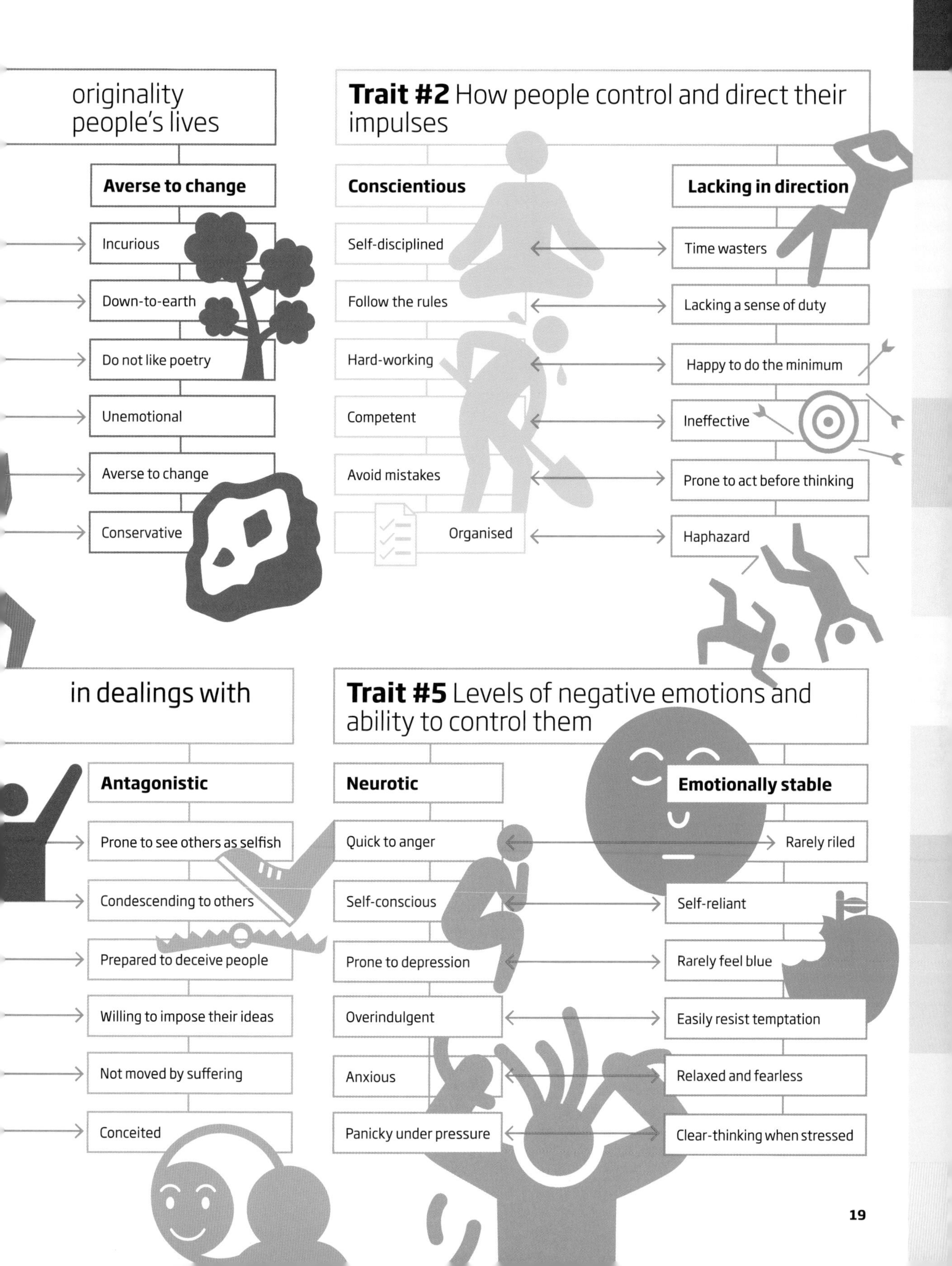
originality
people's lives
Averse to change
Incurious
Down-to-earth
Do not like poetry
Unemotional
Averse to change
Conservative
Trait #2 How people control and direct their impulses
Conscientious
Self-disciplined
Follow the rules
Hard-working
Competent
Avoid mistakes
Organised
Lacking in direction
Time wasters
Lacking a sense of duty
Happy to do the minimum
Ineffective
Prone to act before thinking
Haphazard
in dealings with
Antagonistic
Prone to see others as selfish
Condescending to others
Prepared to deceive people
Willing to impose their ideas
Not moved by suffering
Conceited
Trait #5 Levels of negative emotions and ability to control them
Neurotic
Quick to anger
Self-conscious
Prone to depression
Overindulgent
Anxious
Panicky under pressure
Emotionally stable
Rarely riled
Self-reliant
Rarely feel blue
Easily resist temptation
Relaxed and fearless
Clear-thinking when stressed

What is this thing called personality?

One of the most striking features of life is how differently each of us approaches it. Some people love bright lights and fast cars, others will go to any length to avoid such things. A novel social situation that is a treat for one individual is a cause of sleepless nights for another. Some people fret about saving for their pensions, whereas others spend money as soon as, if not before, they have it. Where do such differences come from?

Circumstances undoubtedly play a part. Some human behaviour is easily explained by social context – for example, people living in a dangerous environment generally think in a more short-term way. Other variations are simply quirks. What interest psychologists most, though, are the systematic ways in which individuals differ, otherwise known as personality traits.

Why do we all have different personalities? Why hasn't natural selection homed in on optimum character traits instead of allowing so much variability? As the study of personality moves onto a more scientific footing, we are starting to understand the underlying neurobiology, and to see that each personality trait is beneficial in certain circumstances and costly in others. We may think that some personality types are more desirable than others, but there is no 'best' personality. It takes all sorts to make a world.

Psychologists think of personality traits as thermostats within the brain, each regulating a range of behaviours and attitudes. Some of these behaviours and attitudes seem to be linked. For example, people who are highly competitive and like loud music and travel tend also to have high sex drives. People who have a specific phobia tend to worry a lot about other things too, and they are more prone to depression. The same people who have trouble resisting the temptation to take drugs have an increased likelihood of developing problems with gambling and antisocial behaviour in general. From such correlations, we infer that there are a limited number of thermostats, each working independently. Your thermostat setting represents where you stand on a continuous scale for each personality trait.

Five key thermostats account for most of the variation in personality. These big five – extroversion, neuroticism, conscientiousness, openness to experience and agreeableness – define five axes along which all individuals fall. Your personality consists of five different scores measured by a personality questionnaire. Since each axis is continuous and they are independent of one another, there are countless unique personality configurations.

How you respond

Where you stand on any one of the big five will show up in a particular type of situation. A person's level of extroversion manifests itself in response to pleasure-seeking or risk-taking activities, with the most introverted people seeming to find these least rewarding. When things get threatening or dangerous, or are perceived to be so, neuroticism is key – highly neurotic people will become anxious or stressed, whereas low scorers won't be so badly affected.

Conscientiousness is all about where your goals are oriented. High scorers stick rigidly to plans or principles while low scorers are more spontaneous. Agreeableness comes to the fore in our personal relationships. Highly agreeable people pay attention to the needs and emotions of others. Low scorers are not so oriented to these cues. Openness determines a person's response to ideas. High

Daniel Nettle is professor of behavioural science at Newcastle University, UK. He is author of *Personality: What Makes You the Way You Are*

scorers like to entertain aesthetic, metaphorical or esoteric ideas; low scorers avoid them.

We are now beginning to relate the big five to the brain. Take neuroticism. Neuroscientists know which parts of the brain are involved in the response to threats: activity centres on structures called the amygdalae. Brain scans reveal that the amygdalae of individuals who score highly in neuroticism have a higher baseline level of metabolic activity than those of low scorers. High scorers also show greater activity in response to distressing stimuli. There is even evidence that the size of the amygdalae is proportional to a person's neuroticism score. Amazingly, the simple, self-rating questionnaires used by personality psychologists actually turn out to measure something about the nervous system that can be verified through objective scientific techniques.

Looking in the mirror

You can roughly measure your own personality by thinking about your typical behaviours. Faced with a potential threat, are you anxious or unflappable? Are you intrigued by new ideas or prefer to stick to what you know? Do you make concrete plans or just let things happen? Are your dealings with others warm or stand-offish? Are you outgoing and enthusiastic or quiet and stoical? If you lean towards the first on any of these, you are likely to score highly on the relevant trait: neuroticism, openness, conscientiousness, agreeableness and extroversion (see 'What are you like?', p 18).

Rewarding activities

We can tell a very similar story for extroversion. A suite of structures in the midbrain is responsive to stimuli that betoken rewards: sweet foods, money, pictures of members of the opposite sex, addictive drugs and so on. These reward centres depend on the neurotransmitter dopamine. There is a linear relationship between a person's extroversion score and their physiological response to a dopamine-like drug called bromocriptine. This strongly suggests that extroversion is the outward manifestation of the responsiveness of the brain's reward system.

Conscientiousness involves the ability to control immediate responses in the service of some longer-term goal or plan. Parts of the prefrontal cortex are implicated. We know this from studies of brain damage to this part of the brain, which can lead to previously conscientious people becoming unable to control their impulses. Brain imaging reveals that people with impulse-control problems have less activity in part of the right prefrontal cortex than others. Also, one study of boys with attention-deficit hyperactivity disorder found they had a smaller volume of this part of the prefrontal cortex than normal.

Of course our thermostat settings alone do not determine our behaviour – that depends on a complex interplay between our brain and our environment – but they do influence it to the extent that if we know someone's personality, we can make a pretty good prediction of how they will respond even in a situation we have never seen them in before. What's more, these settings seem to be quite stable across an individual's lifespan, reflecting the discovery that there is a substantial genetic component to personality. This genetic element also explains why people sharing similar social and cultural backgrounds can be so different.

Genetic impact

Intriguing and sometimes controversial research has implicated several genes in personality differences. For example, the mood-affecting neurotransmitter serotonin is cleared from the gaps between neurons by a protein called serotonin transporter, which can be produced from either of two common forms of the serotonin transporter gene. People with at least one copy of the short form of the gene have higher neuroticism scores than people with two copies of the long form. The genetic findings also correlate with the neuroscientific ones. In the brain scanner, individuals with one or more short forms of the gene show a greater spike of amygdala activity when they are presented with a fearful face than do individuals with two long forms.

We are making progress in understanding the mechanisms that give rise to behaviour, but until recently, relatively little attention had been paid to why evolution left us with such a variety of personalities. Usually, natural selection acts to reduce variation, with the overall genetic variation in a population the result of an interplay between mutation and selection. Mutation keeps adding more variation through random alterations to genes, while selection removes variation through increased mortality or decreased reproduction of individuals with low fitness. In this evolutionary tug of war, selection should win. So why is there so much variation in personality?

Can you change?

The better you know your own personality, the more aware you will be of its costs and benefits. This puts a new spin on the old question of whether you can change your personality. To some extent we can alter who we are, but we might be better off recognising that for almost any personality profile, there is an optimal environment. So if your personality causes you grief, why not try changing the niche you occupy in this complex ecosystem that is modern life?

The key to this conundrum is that natural selection is not always consistent. How this influences the evolution of personality has become clear as researchers begin to consider rudimentary personality traits in wild animals.
For example, birds called great tits exhibit variation in their exploration behaviour. Some individuals inherit a highly exploratory personality and others a more cautious one. If you measure this trait in wild great tits and relate it to their survival, you find that in years when food is scarce, female birds with an exploratory personality are more likely to survive. In these circumstances it pays to disperse further. However, when resources are abundant, cautious females are more likely to survive. With food plentiful, dispersing too far is an unnecessary hazard. (For males this pattern is reversed, reflecting the different survival pressures that they face.)

This illustrates a powerful point: the optimum level of a personality trait depends on the details of the local ecology. When these vary dramatically across space and time, natural selection cannot home in on a single optimum way of being. That explains why the population of great tits includes both exploratory and cautious individuals.

It also explains a discovery about personality traits in sticklebacks. Those that live in waters where they are under threat of predation are consistently aggressive towards each other and bold towards predators. But where predators are lacking, no systematic character traits are detectable. In other words, a personality trait emerges in the unpredictable environment but not in fish for which life is more stable.

Such findings have obvious extensions to humans. If our range of personalities has evolved as a consequence of a changeable world, then we should expect that each trait is advantageous under some circumstances and costly under others.

Double-edged traits

That is what we find. Adults with high scores for extroversion have more sexual partners and do better in economic and career terms on average. But they are also more likely to be hospitalised as a result of an accident or illness, and have less stable family lives. Since they are more likely to divorce, the men often end up not living with their children.

It is tempting to think of extroversion as an unalloyed blessing, but it is not. Being extrovert will attract you to certain types of situation, bringing certain types of life opportunities – you will do well in some settings. But your personality will entail risks, and some alternative pathways might be closed to you. There will also be situations for which your personality is too risk-prone.

We can see this principle at work with another of the big five, agreeableness. Highly agreeable individuals have good social networks and support. This is because they attract and retain friends and allies. In many ancestral environments, not to mention the present day, this would have served them well. However, in devoting time and effort to the needs of others, they incur costs in terms of their own agendas.

Agreeableness is a negative predictor of success, for example, in the worlds of executives and artists, where people need to put themselves first and focus on what they want. What is the optimum level of agreeableness? Apart from it not being at the extremes of the distribution – psychopaths at one end and people with something called 'dependent personality disorder' at the other – there is no correct answer. Whatever your level of agreeableness, there will be benefits and costs.

Likewise, high conscientiousness may help you get the job done, but could also blinker you to other possibilities that might occur to a more easily distracted person, or someone whose approach to life is more flexible. High openness, meanwhile, has been linked with increased social and sexual success. These advantages, however, tend to accrue mostly in historical eras and geopolitical contexts in which artistic types are highly valued. At other times and places, particularly when a population is struggling for survival, practical and capable characters will be much more in demand. Even highly neurotic people can console themselves that when the threat of danger is real, their vigilance may serve them better than the heedless attitude of more laid-back individuals.

Where does morality come from?

Whoever first came up with the idea that humans have an angel on one shoulder and a devil on the other was onto something. We are both the basest of animals and the most noble. Ours is a species capable of cruelty, genocide, war, corruption and greed. Yet we can also be caring, kind, fair and philanthropic. What lies behind this dual nature?

Our capacity for good and evil has exercised philosophers from at least Plato onwards, but today some of the most exciting ideas are coming from evolutionary biologists. They have probed questions of altruism, conscience, prejudice and hatred. Their answers suggest that good and evil are not really all that different.

Pure altruism

Let's start with goodness. The key to virtue is altruism, which is by definition selfless. Genuine altruism is rare in nature. Many animals help others, but only their relatives. In evolutionary terms that is actually a form of selfishness, since it ensures the survival of their own genes.

Yet humans do appear to behave selflessly. Since the 1980s, economists have used games to assess our altruistic tendencies. First came the 'ultimatum game': player A is given some money and told to split it with a second anonymous player B. If B accepts, both keep their share; if not, neither gets anything. It is free money, so A should offer as little as possible and B should accept any amount. But that is not what happens. Instead, player A routinely offers 40 to 50 per cent. Any less and B tends to reject the offer. Even in a version called the 'dictator game', where A can offer 10 or 50 per cent and B has no option to reject, three quarters of people make the more generous offer.

Nobody is under any illusions about this. Pure altruism doesn't make evolutionary sense: there must be something in it for the altruist. Working on this principle, biologists have come up with a variety of explanations.

The first possibility is rather disheartening. Traditional hunter–gatherer groups tend to consist of closely related people. Helpful people are usually promoting their own genetic interests. So we may have evolved strong nepotistic instincts, and niceness to non-relatives is simply overspill.

A more positive explanation is reciprocity – you scratch my back and I'll scratch yours. Humans live in groups and are highly dependent on others. We also remember who owes us a favour. Acts of generosity are actually selfish favours that we can call in when we ourselves are in need.

Altruism may also be driven by the need to maintain a good reputation. Humans are nosy: we like nothing better than to gossip. This is how reputations are made and destroyed. Virtues such as generosity, fairness and conscientiousness are universally valued and people who are seen to display them are often rewarded with material goods and sex. So a good reputation can boost your chances of survival and reproduction.

Altruism could also have evolved to benefit groups. Those that pulled together would have beaten groups whose individuals were more selfish, ensuring their survival. This 'group selection' is a controversial idea, but it is increasingly being accepted as an important driving force behind the evolution of altruism.

Being nice has its downsides, however. It creates an environment where selfish people can enjoy the benefits of cooperative living without paying the costs. To counter this we have evolved a few strategies to discourage free-riders.

One is our seemingly innate desire to punish those who don't pull their weight. People playing the ultimatum game will often reject stingy offers just to see their partner suffer. In the real world we use gossip, censure and ostracism to punish minor misdemeanours and deploy police, courts and prisons to punish more serious crimes.

The right thing to do

Fear of being punished is not the only thing that keeps us in check. Often we are virtuous simply because it feels like the right thing to do. This includes things that are not personally beneficial but will be good for society if everyone does them – things like voting, recycling and giving money to people in need. Such behaviours appear to be driven by a learned association we call 'conscience'.

We learn the social rules of our culture and they become linked in our brains with emotions such as pride and honour, shame and guilt. Selfishness may be in your interest, but it is associated with negative emotions, whereas virtue prompts positive ones.

The pleasure we get from such good deeds is probably induced by a cocktail of neurochemicals including oxytocin. It is normally associated with feel-good activities such as sex and bonding, but is also linked to morality. People with more oxytocin are more generous and caring, and our oxytocin level increases when someone puts trust in us.

Altruism, then, makes us less selfish but, perversely, is also behind some of our most heinous acts. That's because the flip side of generosity to one's in-group is meanness to outsiders. This is underpinned by oxytocin, too, and is sometimes called the 'mama bear effect' because it mirrors the urge to defend cubs at any cost. As a result, altruism can promote atrocities such as racism, genocide and war. Just as almost everyone is capable of altruism, under different circumstances we are also capable of evil, from bullying and corruption to torture and terrorism.

The upshot of this is that good and evil are two sides of the same coin. Evolution has made us both, and we cannot have one without the other.

Blood suckers

Finding true altruism in other species has proved tough. Our closest living evolutionary relatives, chimpanzees, are basically selfish, though they have been found to spontaneously share food. Further afield, one of the only species that routinely acts selflessly is the vampire bat, which will regurgitate blood for a hungry neighbour in the roost even if they are totally unrelated.

This is probably explained as reciprocal altruism – you scratch my back and I'll scratch yours (eventually). Vampire bats don't always find a victim to suck blood from and can starve to death after a couple of nights fasting, so sharing with a roost-mate that is likely to return the favour another night is an obvious strategy to get through lean periods.

Angels or demons?
We are a caring species, capable of kindness and generosity with no expectation of reward. Yet we can also be hostile and cruel. Perhaps nowhere is this dual nature clearer than in the plight of Middle Eastern refugees. Here, people help one another cross the no-man's-land between Macedonia and Serbia. Most are escaping the wars in Syria and Iraq to seek a better life in the European Union, but must contend with fences, border guards and xenophobia. Resolving conflicts like this is not our forte. We are hampered by our tribal nature: for example, the hormone that makes us kind to people like us also makes us hostile to outsiders. Can we learn from this knowledge? The playwright Anton Chekhov thought so: humans, he said, will only become better when you make them see what they are like.

Credit: Rocco Rorandelli/ TerraProject

Why are humans so generous?

Life isn't easy as a Maasai herder. At any moment, disease could sweep through your livestock, the source of almost all your wealth. Drought could parch your pastures, or bandits could steal the herd. No matter how careful you are, or how hard you work, fate could leave you destitute. What's a herder to do?

The answer is simple: ask for help. Thanks to a Maasai tradition known as *osotua* – literally, umbilical cord – anyone in need can request aid from anyone else. Anyone who's asked is obliged to help, often by giving livestock, as long as it doesn't jeopardise their own survival.

Each individual maintains a network of osotua partners. Once formed, the relationships last for generations, with children inheriting them when their parents die. But no one expects a recipient to repay, and no one keeps track of how often people ask or give.

Osotua runs counter to the way we usually view cooperation, which is all about reciprocity – you scratch my back, I'll scratch yours. Yet similar forms of generosity turn out to be common in cultures around the world.

That's a curious discovery. In biological and evolutionary terms, it makes no sense to give and get nothing in return. Altruism is one thing – it usually promotes the altruist's social standing or biological fitness in some way – and is common in nature. But pure generosity is pretty much unique to humans. Are we generous by nature? How did we get to be this way? In short, what does it take to make the milk of human kindness flow?

Osotua isn't confined to the Maasai. In every society anthropologists have studied, they find examples of generosity that is based on need. Fijians, Tanzanian slum dwellers, American cattle ranchers and even Western urbanites all pitch in to help people in need, with no expectation of being paid back.

The giving is often extremely one-sided. Among the Maasai, the same wealthy families are approached again and again. On the face of it, they seem to lose much and gain little by participating. Why do they continue to put their hands in their pockets against their apparent best interests?

Unpredictable events

A clue lies in the trigger for osotua: an unpredictable crisis. This suggests that these practices persist because they help manage risk, which pays off for everyone in the long run. Even the best-prepared family can fall prey to catastrophe such as a sudden illness. These types of risk cannot be prevented, so need-based giving may have emerged as a proto-insurance policy.

Prosperous members of many societies share so that this social insurance will be available if they need it – just as wealthy homeowners insure their belongings against the small risk of them going up in smoke. Need-based giving works best when risks are 'asynchronous' – when hardship is likely to strike one family and spare their neighbours.

Herding tribes in northern Mongolia, for example, use such generosity to help families struck by illness. However, the system breaks down when they face their biggest threat – a severe winter or '*zud*' which starves or freezes millions of animals to death. With everyone affected, helping one's neighbours isn't an option.

Yet, the ability to help isn't enough in itself. To benefit from Maasai-style generosity, you need to prevent cheating; for example, asking when not truly in need. In some societies the solution is easy. It's hard to hide livestock, so herders in Mongolia can't pull the wool over their neighbours' eyes.

In addition, requests tend to be made in public, so everyone knows who has asked and given – or refused to give.

Where wealth is easier to hide, reputation is the key. In Fiji, for example, there is an osotua-like practice called *kerekere*. People who ask repeatedly get a reputation for shameful laziness, which makes people think carefully before making kerekere requests.

In fact, reputation appears to be the rock upon which generosity is built. When anthropologists gave Fijian men a sum of money roughly equal to a day's wages, and the choice of sharing their windfall with people they knew, they proved surprisingly generous. On average, they kept just 12 per cent of the money for themselves, and about half the men gave it all away.

When asked how they chose who to share their money with, almost all said they gave to people who needed it. However, closer statistical analysis showed that reputation was almost as important as need. Men who had a reputation for giving tended to be the ones who received more. In day-to-day life, norms of generosity, love and respect drive decisions about sharing more than cold cost–benefit calculations do. This reinforces the idea that generosity is good. But do humans become less generous when they live in more complex societies?

Bah, humbug

The Ik of Uganda were once described as the least generous people in the world. They acquired this reputation from an anthropologist called Colin Turnbull who lived among them in the 1960s and later described them as 'unfriendly, uncharitable, inhospitable and generally mean as any people can be.' He was especially horrified at their selfishness, refusing to share food with starving relatives – even their own children. But the Ik were misunderstood. They had recently been displaced to make way for a national park and were suffering severe hardship. We now know they are just as capable of extraordinary acts of generosity as any other humans.

Choosing who to help

People in Western societies often walk past beggars on the street. But that could be because they know that social institutions exist, and expect them to step in and help.

In fact, Westerners often give generously to strangers. When natural disasters occur, often in distant lands, people donate to charity without any expectation of return.

Arguably, that is even more generous than a system like osotua. People living in smaller-scale societies tend to direct their generosity towards people they know. Fijians, for example, are very generous within their village. But when anthropologists ask them about giving to distant poor people, they seem baffled as to why anyone would send money to someone they don't know far away. It seems generosity runs deep in our species, perhaps even to a fault.

Why do we fight?

Consider yourself lucky. You are living in the most peaceful era of our species' existence. Today, you are less likely to die at the hands of another person than at any time in human history.

If you had lived in a prehistoric society, you would have had about a 1 in 200 chance per year of being murdered or killed in conflict. In modern Western societies, that number is now 1 in 100,000. According to Harvard psychologist Steven Pinker in his exhaustive history of human violence, *The Better Angels of Our Nature*, violent deaths – from revenge killing and blood feuds to genocides and wars – have been declining for at least the past 6,000 years.

This timeframe is too short for the decline to be put down to evolutionary changes. Our propensity for violence is still lurking in the background, but has been dialled down by shifts in culture – changes in politics, law, morality and an increased cosmopolitanism that has allowed people to vicariously experience and empathise with others around the globe.

Despite this, deadly violence is still an aspect of human existence. About two million people a year still die as a result of homicide or war, which is around 3 per cent of all fatalities.

Redder in tooth and claw

By the standards of other animals, humans are not especially violent. Nature is still redder than us in tooth and claw. But when it comes to reasons why we fight, humans are highly irregular. Other animals fight over limited resources and desirable mates, and will often gang up to do so. Wolves from one pack will team up to take out members of another and chimpanzee troops sometimes fight their neighbours in ways that resemble human conflict. Between 1974 and 1978, two troops of chimpanzees in Gombe Stream National Park in Tanzania fought a protracted conflict that primatologist Jane Goodall later described as a 'war'.

Humans wage turf wars too, but we also fight over more intangible things such as honour, values, ideas, beliefs and symbols of cultural identity.

Bands of brothers

Our innate tribalism sometimes leads to something called 'fusion', where individual identities are subsumed by the group. An effective way to induce this is ritual: synchronised activities, from liturgical recitation to military goose-stepping, seem to make people more likely to follow orders to be aggressive to others.

Rituals that produce shared suffering, pain and fear are especially good at catalysing fusion, which explains the bizarre and often dangerous initiation ceremonies seen in warrior cultures and university drinking clubs. Intense and terrifying shared experiences have a similar effect. During warfare, groups of soldiers often fuse into 'bands of brothers' who are willing to die for one another even if they don't believe in the cause they are fighting for.

This type of conflict seems to be an integral part of human nature. Even though we're one species, we cannot help but define ourselves as belonging to multiple smaller groups. Nationality, ethnicity, religion, political party, civic pride and even the football teams we support are some of the things that divide us into mutually hostile 'in-groups'. The boundaries of these groups are fluid and they can fuse or schism depending on circumstances. Fans of rival football clubs can forget their quarrels when supporting the national team.

This tribal antagonism has its root in our distant past when our ancestors lived in small roaming bands of (mostly) related individuals in frequent conflict – and occasional alliance – with neighbours. Archaeological evidence going back 12,000 years and studies of tribal peoples suggest that around 15 per cent of all deaths in these societies stemmed from inter-group violence. Tribes made up of individuals prepared to fight for a common good had a competitive edge over those that weren't, and so tribal violence was selected by evolution. Unfortunately we have dragged its baggage into the modern world.

Tribalism and the discord it engenders are frighteningly easy to induce, as social psychologists have long been aware. More than 40 years ago, the late Henri Tajfel showed that dividing a group of strangers into two teams based on arbitrary criteria such as whether they preferred the paintings of Klee or Kandinsky triggered their tribal instincts. Members of the Kandinsky tribe behaved favourably towards team-mates while treating members of the other team harshly, and vice versa. Since then, many experiments have revealed how the flimsiest and most transient badges of identity can trigger people to divide themselves into 'us' and 'them' – even the colour of T-shirts randomly assigned by psychology researchers can do it.

Paradoxically, these antagonistic tendencies may be driven by a much more noble side of human nature: our unparalleled capacity for large-scale cooperation and altruism. Few activities draw on these traits like fighting on behalf of our group. Love for one's own group could easily have co-evolved with hostility towards outsiders, creating an unusual and incendiary blend of mutual support and hostility (see 'Where does morality come from?', p 24).

Sacred values

Group identity is reinforced by culture, which encourages members to differentiate themselves from others through markers such as dress codes, food preferences and rituals. It also prescribes what is worth fighting for.

The most powerful motivators of all are what psychologists call 'sacred values' – beliefs shared by all group members that cannot be traded for material things like food or money. Offering people such rewards to go against their sacred values doesn't just fail, it usually backfires, leading to moral outrage and even stronger commitment to the value. Sacred values are absolute and non-negotiable, which is why they loom large in many contemporary conflicts.

As their name implies, sacred values are often religious in nature, but they don't have to be. Freedom of speech, liberty, democracy and the stewardship of nature are considered sacred values by some. Membership of the group per se may be a sacred value. Whatever they are, and however noble or ignoble, we will fight tooth and nail for them.

Violence is often seen as a throwback to our animal past. But in some ways, there is nothing more human.

The strange nature of belief

'Alice laughed. "There's no use trying," she said: "one can't believe impossible things."

'"I daresay you haven't had much practice," said the Queen. "When I was your age, I always did it for half-an-hour a day. Why, sometimes I've believed as many as six impossible things before breakfast."'In Lewis Carroll's day, believing impossible things would more than likely have been seen as a sign of mental imbalance. Today, we know that it is quite normal. Six before breakfast is probably about par for the course.

Guidebook to reality

Whatever beliefs you hold, it's hard to imagine life without them. Beliefs come in all shapes and sizes, from the trivial (I believe it will rain today) to the profound and life-shaping (I believe in God / socialism / extrasensory perception). But however big or small, together they form a personal guidebook to reality, telling us not just what is factually correct but also what is right and good, and hence how to behave towards one another and the world. They also come so naturally to us that we rarely stop to think where they actually come from.

The common-sense assumption is that we reason our way to our beliefs, weighing up facts and evidence and coming to rational and defensible positions. But the more we discover about belief, the more naive that looks. When it comes to what we believe and why, it turns out we have a lot less control than we might think.

Scientists once thought human beliefs were too complex to study. Not any more. What is emerging is a picture of belief formation that conspicuously excludes reasoned examination of the available evidence. Instead, our beliefs come from three main sources – our evolved psychology, personal biological differences and the company we keep.

The importance of evolved psychology is illuminated by perhaps the most important belief system of all: religion. Although the specifics vary, religious belief per se is remarkably similar across the board. Most religions feature a familiar cast of characters: supernatural agents, an afterlife, moral directives and answers to existential questions. Why do so many people believe such things so effortlessly?

According to the 'cognitive by-product' theory of religion, their intuitive rightness springs from basic features of human cognition that evolved for other reasons. In particular, we tend to see patterns everywhere and assume that agents cause events. Both of these features were useful survival strategies for our distant ancestors, and have produced a brain that is primed to see agency and purpose everywhere. Both are important features of religion – particularly the idea of an omnipotent but invisible agent that makes things happen and gives meaning to otherwise random events. Humans are naturally receptive to religious claims, and when we first encounter them we accept them unquestioningly.

Politics and biology

A second source of rightness is more personal. Political belief is often rooted in our basic biology – especially emotions such as fear and disgust. For example, conservatives generally react more fearfully than liberals to threatening images, and are more disgusted by stimuli like fart smells and rubbish. Conservatives are biologically predisposed to see the world as threatening and unclean, and to act accordingly. These innate differences may lie at the root of differences in opinions over issues such as law and order, national security, same-sex marriage and immigration.

Curious convictions

Even the most normal people believe in the strangest things. About half of US adults endorse at least one conspiracy theory. Belief in paranormal or supernatural phenomena is widespread, and superstition and magical thinking are nearly universal.

Surprisingly large numbers of people also hold beliefs that a psychiatrist would class as borderline delusional. These are mild versions of beliefs that could get you diagnosed with a mental illness – that your family has been abducted and replaced by impostors, for example, or that a celebrity is sending you secret messages. Most of us seem untroubled by them in our everyday lives.

Least surprisingly, belief is powerfully shaped by the culture we grow up in. We are social beings and many of our beliefs are learned from the people we are closest to.

But this is often more about belonging than about being right. Our strong social nature means that we adopt beliefs as badges of cultural identity. This is often seen with hot-potato issues, where being a member of the right tribe can be more important than being on the right side of the evidence. Acceptance of climate change, for example, has become a shibboleth in the US – conservatives on one side, liberals on the other. Evolution, vaccination and others are similarly divisive issues.

Building on sand

Wherever our beliefs originate, by the time we reach adulthood we tend to have formed a fairly stable, internally consistent belief system that stays with us for the rest of our lives. But the idea that this is the product of rational, conscious choices is highly debatable. It also leaves room for unsupported assumptions, prejudices, contradictions, superstitions and all manner of other impossible things – not just before breakfast but all day, every day.

The upshot of all this is that our personal guidebook is both built on sand and also highly resistant to being shaken. We will go to extraordinary lengths to reject information that contradicts our existing beliefs, or seek out further information to reaffirm what we already believe. This 'confirmation bias' is a universal human trait that often blinds us to the true nature of our own beliefs.

That's not to say that beliefs cannot change. Presented with enough contradictory information, we can and do change our minds. Often, though, rationality isn't the main driver. We are more likely to change our beliefs in response to a compelling moral argument or a new social circle – and when we do change, we reshape the facts to fit with our new belief.

The world would be a boring place if we all believed the same things. But it would surely be a better one if we all stopped believing in our beliefs quite so strongly.

Why religion is as natural to us as talking

By the time he was five years old, Wolfgang Amadeus Mozart could play the clavier and had begun to compose his own music. Mozart was a 'born musician'; he had strong natural talents and required only minimal exposure to music to become fluent.

Few of us are quite so lucky. Music usually has to be drummed into us. And yet in other domains, such as language or walking, virtually everyone is a natural; we are all 'born speakers' and 'born walkers'.

So what about religion? Is it more like music or language?

Naturally inclined

Research in psychology, anthropology and the cognitive science of religion tells us that religion comes nearly as naturally to us as language. The vast majority of humans are 'born believers', naturally inclined to find religious claims and explanations attractive and effortlessly acquirable, to attain fluency in using them, and to continue to use them throughout life.

This attraction to religion is an evolutionary by-product of our ordinary cognitive equipment, and while it tells us nothing about the truth or otherwise of religious claims, it does help us to understand the near-ubiquity of religious belief.

As soon as they are born, babies start to try to make sense of the world around them. As they do so, their minds show regular tendencies.

One of the most important is to recognise the difference between ordinary physical objects and 'agents' – things that can act upon their surroundings. Even babies know that balls and books must be contacted in order to move, but agents such as people and animals can move by themselves.

Because of our social nature we pay special attention to agents and are strongly attracted to explanations of events in terms of their actions – particularly events that are not readily explained by ordinary causation.

In the first year of life, for example, children intuitively grasp that apparent order and design, such as we see in the world around us, requires an agent to bring it about and that while agents can create order or disorder, non-agents such as storms only create disorder.

Babies also seem sensitive to another feature of agents: they need not be visible. Babies do not need a person, an animal, or anything at all to get their agency reasoning up and running. This is an important skill if they are going to apply their reasoning about agents to function in social groups, avoid predators and capture prey. All require us to be able to think about agents we cannot see. This hair-trigger agent reasoning and a natural propensity to look for agents in the world around us – even ones we cannot see – are part of the building blocks for belief in gods. Once coupled with some other cognitive tendencies and ordinary learning strategies, they make children highly receptive to religion.

Searching for purpose

One of those tendencies is the search for purpose. From childhood we are very attracted to the notion that everything has a purpose – including animals and people, trees and icebergs. Four- and five-year-olds think it more sensible that a tiger was 'made for eating and walking and being seen at the zoo' than that 'though it can eat and walk and be seen at the zoo, that's not what it's made for'.

When it comes to speculation about the origins of natural things, children are receptive

Justin L. Barrett is professor of psychology at Fuller Theological Seminary in Pasadena, California, and author of *Why Would Anyone Believe in God?*

to explanations that invoke design or purpose. It seems more sensible to them that animals and plants were brought about for a reason than that they arose for no reason. Children tend to embrace creationist explanations of living things over evolutionary ones. Adults do not outgrow this attraction but must have it drummed out of them through formal education.

Of course gods do not just create or order the world, they typically possess superpowers: superknowledge, superperception and immortality. These properties of gods also turn out to be intuitively easy for children to adopt.

A God-shaped space

Even though children are naturally receptive to religious ideas, this concept of religion deviates from theological beliefs. Children are born believers not of Christianity, Islam or any other theology but of what I call 'natural religion'. They have strong natural tendencies toward religion, but these tendencies do not inevitably propel them towards any one religious belief.

Instead, the way our minds solve problems generates a God-shaped conceptual space waiting to be filled by the details of the religious culture into which they are born.

Children appear to presume that all agents have superknowledge, superperception and immortality until they learn otherwise. They find it easier to assume that others know, sense and remember everything than to figure out precisely who knows, senses and remembers what. Their default position is to assume superpowers until teaching or experience tells them otherwise.

Mind-reading

This assumption is related to the development of a faculty called 'theory of mind', which concerns our understanding of others' thoughts, wants and feelings. Theory of mind is important for social functioning but it takes time to develop. Some three- and four-year-olds simply assume that other people have complete, accurate knowledge of the world (at least as they understand it).

A similar pattern is seen with children's understanding of the inevitability of death. The default assumption is that others are immortal.

These various features of developing minds make children naturally receptive to the idea that there may be one or more gods helping to account for the world around them.

When they encounter religious ideas children find them intuitively plausible and gravitate towards them, often holding on to them for life. These ideas supply an easy-to-understand account for perceived order and purpose in nature, for great fortune and misfortune, and for matters that concern morality, life, death and the afterlife.

Importantly, these ideas do not need to be drummed in to children. We do not need to be indoctrinated to believe in gods. Religious belief is the default path of the human mind.

2 The Self

moving spot changing colour. This edited version of the action is then screened in the theatre of consciousness.

Unfortunately, this explanation does not fit well with evidence of how perception works. Conscious responses to visual stimuli can occur at a speed very close to the minimum time required for information to reach the brain and be processed. If we add up the time these take, there is not enough time left for a delay of sufficient length to explain the moving-spot effect.

Mistaken perception

Perhaps there is something wrong with the notion of a self perceiving a stream of unified sensory information. Perhaps there are just various neurological processes taking place in the brain, without some central agency where it all comes together. This makes it much easier to make sense of the moving-spot illusion.

The perception of a green spot turning red arises in the brain only *after* the perception of both spots. Our mistaken perception of the real flow of events is akin to the way we interpret the following sentence: 'The man ran out of the house, after he had kissed his wife.'

The sequence in which the information comes in from the page is 'running–kissing', but the sequence of events you construct and understand is the opposite: 'kissing–running'. For us to experience events as happening in a specific order, it is not necessary that information about them enters our brain in that same order. The brain sorts it out after the fact.

That view is supported by another odd perceptual phenomenon called the 'flash-lag illusion'. You watch a rotating disc with an arrow on it. Next to the disc is a spotlight that is

Broken selves

Your sense of self can be disturbed by many things. Depersonalisation disorder, for example, is a mental illness defined by a persistent feeling of detachment. It has been described as feeling like living in a dream, or being an external observer of your life. Psychedelic drugs such as LSD create a similar sensation. Alongside sensory distortions, a common experience is a feeling that the boundary between one's self and the rest of the world is dissolving. Of all the disturbances of the self, the eeriest and least understood is Cotard's syndrome. Symptoms range from claims that internal organs have gone missing to a belief that one has ceased to exist. People with the delusion have been known to arrange their own funerals.

programmed to flash at the exact moment the arrow points to it. Yet this is not what you perceive. Instead, the flash apparently occurs after the arrow has passed.

One explanation is that the brain is projecting into the future. Visual stimuli take time to process, so the brain compensates by 'seeing' the moving arrow where it will be in a few moments' time. The static flash – which it can't anticipate – seems to lag behind.

But that cannot be right. If the disc stops rotating at the exact moment the arrow passes the light, the illusion disappears. If the brain were predicting the future, it would persist. What's more, if the arrow starts stationary and moves in either direction immediately after the flash, the movement is perceived before the flash. How can the brain predict movement that doesn't start until after the flash?

The explanation is that the brain is assembling a plausible story of what happened retrospectively. The perception of what is happening at the moment of the flash is determined by what happens to the arrow after it. This seems paradoxical, but other tests have confirmed that what is perceived as 'now' can be influenced by what happens later.

Who's in control?

The third and final core belief is that the self is the locus of control. Yet cognitive science has shown in numerous cases that our mind can conjure, post hoc, an intention for an action that was not brought about by us.

In one experiment, for example, volunteers are asked to move a cursor slowly around a screen on which 50 small objects are displayed, and asked to stop the cursor on an object every 30 seconds or so. The computer mouse controlling the cursor is shared, Ouija-board style, with another volunteer. Via headphones, the first volunteer hears words, some of which relate to the objects on screen. What this volunteer does not know is that their partner is one of the researchers, who occasionally forces the cursor towards a picture.

If the cursor is forced to the image of a rose, and the volunteer had heard the word 'rose' a few seconds before, they report feeling that they had intentionally moved the cursor there. The reasons why these cues combine to produce this effect is not what is interesting here, though. More important is that it reveals one way in which the brain does not always display its actual operations to us. Instead, it produces a post-hoc narrative for the cursor's movement despite lacking any factual basis for it.

Challenging our core beliefs

So, many of our core beliefs about our selves do not withstand scrutiny. This presents a huge challenge for our everyday view of what we are, as it suggests that in a very fundamental sense we are not real. Instead, our self is comparable to an illusion – but without anybody there to experience it . . .

Ultimately, though, we may have no choice but to endorse these mistaken beliefs. Our whole way of living relies on the notion that we are unchanging, coherent and autonomous individuals. The self is not only a useful illusion; it may also be a necessary one.

Building the self from scratch

What is the self? René Descartes encapsulated one idea of it when he wrote 'I think, therefore I am.' He saw his self as a constant, the essence of his being, on which his knowledge of everything else was built. Others have very different views. A century later, philosopher David Hume argued that there was no 'simple and continued' self, just the flow of experience. Hume's proposal resonates with the Buddhist concept *anattā*, or non-self, which contends that the idea of an unchanging self is an illusion.

Today, a growing number of philosophers and psychologists agree that the self is an illusion. But there is still much to explain. For example: how you distinguish your body from the rest of the world; why you experience the world from a specific perspective, typically somewhere in the middle of your head; how you remember yourself in the past or imagine yourself in the future; and how you are able to conceive of the world from another's point of view. Science is close to answering many of these questions.

No longer an essence

A key insight is that the self should be considered not as an essence, but as a set of processes like programs running on a computer. Some patterns of brain activity constitute processes that generate the self. This fits with Hume's intuition that if you stop thinking, the self vanishes. For instance, when you fall asleep, 'you' cease to exist. However, when you awake, those same processes pick up much where they left off.

The idea that the self emerges from a set of processes has encouraged the belief that it can be recreated in a robot. By deconstructing it and then attempting to build it up again piece by piece, we might learn more about what selfhood is, and perhaps even resolve its central mystery: why it feels compellingly real, yet, when examined closely, seems to dissolve away.

How, then, might we deconstruct selfhood in order to build it in a machine? Philosophy, psychology and neuroscience give many insights into what constitutes the human self.

William James, a founder of modern psychology,

Your five selves

Psychologist Ulric Neisser's influential theory of the self breaks it down into five elements:

- The ecological self is what distinguishes you from others, gives you a sense of body ownership and an individual point of view.
- The interpersonal self underlies self-recognition (e.g. in a mirror), and allows you to see others as agents like you and have empathy for them.
- The temporally extended self endows you with awareness of your personal past and future.
- The conceptual self is the idea of who you are: a being with a life story, personal goals, motivations and values.
- The private self is your inner life: your stream of consciousness and awareness of it.

Tony Prescott is professor of cognitive neuroscience at the University of Sheffield, UK

suggested that the self can be divided into two: an 'I' that comprises the experience of being and a 'me' that is the set of ideas you have about your self. Then, in the 1990s, Ulric Neisser, a pioneer of cognitive psychology, went further. He identified five key aspects of self: the ecological self, the interpersonal self, the temporally extended self, the conceptual self and the private self. Neisser's analysis is not the final word, but it provides useful guidance about what might be required to build an artificial self.

Body schema

Say we want to emulate the ecological self. Key to this is an awareness of one's body and how it interacts with the world. To achieve this, a robot would need an internal 'body schema' – a process that maintains a model of its physical parts and its body pose, and which distinguishes its physical self from others.

Then there's the temporally extended self, which is an awareness of one's continuity in time. Insight about this can be found in the case of a man known as N.N., who lost the ability to form long-term memories after he had been in an accident. The damage to his brain also left him without foresight. He described trying to imagine his future as 'like swimming in the middle of a lake. There is nothing there to hold you up or do anything with.' In losing his past, N.N. had also lost his future. Brain-imaging studies have since confirmed that the same brain systems that underlie our ability to recall past events also allow us to imagine the future.

Next is the interpersonal self. A key element of this is empathy, which derives from a general ability to imagine oneself in another's shoes. One way we might do this is to employ the ecological self to internally simulate what we perceive to be another's situation.

So the interpersonal self is related to the ecological self. But there is more. Another important building block may be the capacity to learn by imitation. Your ability to interpret another person's actions using your own body schema is partly down to mirror neurons – cells in your brain that fire when you perform a given movement and when you see someone else perform it.

We and others have made a start on modelling the ecological, interpersonal and temporal selves in robots, but our efforts are undoubtedly crude in comparison to what goes on in human brains, and there is a long way to go. We have yet to tackle the conceptual and private selves.

Origins of personhood

There are also ethical issues to consider. Are there aspects of the human self that should not be emulated in robots, such as motivations and goals? If we created a robot with a sense of self, would we also have to grant it personhood?

You might also argue that this all misses a crucial element: the 'I' at the centre of James's notion of self, or what we call 'consciousness'. One possibility is that this arises when the other aspects of self are brought together. In other words, it may be an emergent property rather than a distinct thing in itself. Returning to the Buddhist idea, when you strip away the different component processes, perhaps there will be nothing left.

Eternal consciousness
What happens to your mind when you die? Bina Rothblatt hopes hers will live on forever in a sentient robot called BINA48, here seen talking to its caretaker, Nick Meyer of the not-for-profit Terasem Foundation. The real Bina recorded more than a hundred hours of memories, feelings and beliefs which have now been stored in digital form inside BINA48's artificial intelligence. BINA48 can see, connect to the internet, recognise speech, make facial expressions and conduct conversations. Once this artificial consciousness has been perfected, the next step will be to upload it into a body – biological or technological – to create the kind of life experiences that a human would enjoy.

Credit: Max Aguilera Hellweg/ INSTITUTE

Do you have free will?

Did you freely decide to read this page, or did something or someone make you do it? Humans have been wrestling with questions of free will for millennia. Are we truly in control or does some external agent – be that the laws of physics or an omnipotent god – predetermine the trajectory of our lives?

There are no easy answers. Most of us feel very strongly that we have free will – that we can, up to a point, do what we want, when we want. Society is built on this assumption too. We reward people for good deeds and punish wrongdoers. Unfortunately, generations of philosophers and scientists have begged to differ. Like it or not, free will is probably an illusion.

Varying views

From a hard-nosed philosophical perspective, free will is impossible. A choice is not free unless it is uncaused; that is, unless the 'will' is exercised independently of all other external influences. The problem with this is that choices are made by brains, and brains are physical objects that operate according to the rules of causality. The state your brain is in now determines the state it will shift to next.

The evidence from neuroscience is also hard to reconcile with the existence of free will. In a classic experiment in 1982, psychologist Benjamin Libet asked volunteers to sit for a while and then, at entirely their own volition, move a finger. From recordings of their brain activity he discovered a neurological signal that occurred about 550 milliseconds before they moved their finger and, bizarrely, about 350 milliseconds before they even became aware that they were going to. Libet interpreted this to mean that their unconscious brains were preparing to move before the volunteers were aware of it. In other words, free will does not really exist.

Physics is also no friend of free will. Newton imagined the universe to be like a giant clockwork machine running according to immutable laws of motion. The initial conditions determine how it runs, down to the last detail. Such a fully determined universe has no room for free will. Given enough information about its present state, we could extrapolate to any past or future state with 100 per cent accuracy. Everything that has or will happen was determined at the big bang.

Even when Newtonian gravity was superseded by Einstein's theory of general relativity, nothing changed as far as determinism was concerned. According to Einstein, the universe actually exists all at once, and everything that has happened and will happen is already there in what we now call the 'block universe'. Einstein said that any change with passage of time is merely 'an illusion, albeit a persistent one'. This is determinism in all its glory.

Can quantum theory help?

Maybe quantum physics offers us a way out. Quantum theory has certainly changed the picture dramatically by making the universe fundamentally random. When a quantum particle, such as a photon, encounters a piece of glass, such as your window, its behaviour is not determined by any former state. There is a chance that it will go through, but another chance that it will be reflected. As far as we can tell there is nothing in the universe that determines which will happen. It is genuinely random.

At first glance this might seem to create room for free will. What happens in the universe can't be entirely determined from beginning to end because

What if it doesn't exist?

If free will is an illusion, it is a persistent one. Our intuitive certainty in it can be shaken when we are presented with evidence to the contrary, but only briefly.

In experiments, people behave more selfishly and dishonestly if they are persuaded that free will doesn't exist. They are also more likely to treat wrongdoers leniently, offering a hypothetical criminal a shorter prison sentence than they would otherwise have done. But these behavioural changes only last until the powerful feeling of our own agency reasserts itself.

So proving conclusively that there is no free will could have some curious consequences. While people would come to think differently about free will at an abstract level, it's unlikely to have a big impact on the way people actually behave.

you can never know what's going to happen at the quantum scale. But it doesn't actually help. If the universe – including your brain – is fundamentally random, how can you ever be said to have freely chosen to do anything?

That's not the whole story, however. There is an interpretation of quantum mechanics according to which both determinism and randomness can be maintained. According to the 'many worlds' interpretation, all the alternatives exist at the same time – just in different worlds. So in one world the photon goes through your window, while in another parallel world it is reflected. There is a copy of you in each of these parallel worlds.

But this doesn't really help. Not only is it fully deterministic – everything that can happen, does – but you cannot decide which particular world you yourself will occupy. Free will vanishes again.

Walking the line

In the end, it is clear that neither determinism nor randomness is good for free will. If nature is fundamentally random, then outcomes of our actions are also completely beyond our control: randomness is just as bad as determinism.

The paradox deepens ever further when we think about what kind of free will we would like. To be worth having, free will must allow us to choose our actions, but these actions must produce deterministic (that is, non-random) effects. Free will therefore involves walking a fine line between determinism and randomness.

If free will really does not exist, that would be deeply unsettling. The sense of being able to choose one course of action over another is an essential part of being human. If everything is preordained, why bother striving for success, being nice to people or obeying the law?

But that reaction instantly negates itself. If free will does not exist, then you cannot decide to stop acting in a certain way. If it does, then striving for success remains the right thing to do. Physics and neuroscience may take away free will, but they will never take our freedom.

Why there's more to you than nature and nurture

What a dull world it would be if we were all the same. So it's a good job we're all different – and not just in physical appearance but also in personality, intelligence, beliefs, sense of humour and other characteristics. What is it that makes us turn out the way we do?

There's an obvious answer: your mum and dad. Like it or not, who you are owes a great deal to them. As the poet Philip Larkin observed, they f**k you up even if they don't mean to. The old curmudgeon somehow forgot to mention all the good stuff too.

But how? For more than a century, debate has centred on the idea that we are the product of two competing forces, nature and nurture. Nature (or genes) is what we are born with; nurture (or environment) is what influences us during our lives. Most traits are a mixture of both, to varying degrees. But which one is more important?

Consider traits like intelligence or gregariousness. Both could plausibly be a product of nature, or nurture, or both. Do intelligent parents have intelligent children because they fill them full of intelligent genes, or because they fill their houses with books and their family dinners with intelligent conversation? Equally, are gregarious children born or made?

Twin studies

Teasing apart the relative contributions is not easy. The standard way is by looking at pairs of twins. Today, more than 1.5 million twins around the world are enrolled in studies aiming to assess the relative roles of genes and the environment in everything from ageing to religious belief.

Twin studies rest on some simple assumptions. Twins share the same environment in the womb and usually throughout childhood too. Sometimes they also share genes: identical twins are genetically the same, but non-identical twins are not. So, the thinking goes, if identical twins tend to share a trait to a greater extent than non-identical twins do, it is probably largely genetic. If they don't, it is likely to be due to the environment. Rare cases of identical twins separated at birth

Evolved randomness?

Epigenetics may be a way for evolution to hedge its bets. Within our genome, there are hundreds of regions where epigenetic patterns appear totally random – they are neither genetically predetermined (nature) nor set by the environment (nurture), and they vary widely from individual to individual. These regions include many key developmental genes. One possible explanation is that the randomness is an evolved feature. Many animals have to survive in a constantly changing environment. Random epigenetic changes produce lots of variation in genetically similar offspring, increasing the chances that some of them will survive.

and brought up apart can also add useful information, because they are genetically identical but experience different environments as they grow up.

As a result of twin studies we now know a great deal about the relative heritability of various complex traits. So, for example, hair colour is largely genetic and native language environmental. IQ scores are about half-and-half.

In recent years, however, it has become clear that the nature/nurture dichotomy is a false one. It is no longer a case of nature *or* nurture but nature *and* nurture. Genes and the environment work together, often in mutually reinforcing ways. Nowhere is this clearer than in the burgeoning field of epigenetics.

Altered activity

Epigenetic markers are chemical tags added to DNA that alter the activity of genes without altering their genetic sequence. They are added to (and removed from) DNA throughout life in response to environmental factors such as diet, stress and pollution.

The role of epigenetics has been illuminated by twin studies. Despite sharing genes and environment, identical twins sometimes differ markedly in personality, disease susceptibility, even appearance.

Many of these differences have been linked to epigenetic markers. Of particular interest are identical twins where one has a disease and the other does not. For a wide range of disorders including cancer, rheumatoid arthritis and autism, researchers have found different epigenetic profiles in the twins.

Differences have also been linked to behaviour. In one pair of identical twin sisters, for example, different epigenetic patterns were found on a gene implicated in stress and anxiety. That may go some way to explaining their very different career choices: one is a war reporter, the other an office manager.

Of course, epigenetic markers could be seen as being just another form of nurture. But the fact that they are etched onto the genome means they are also nature. Our epigenetic profiles are shaped by the environment, which in turn influences the activity of our genes, which in turn shapes our behaviour, and so on in a complex interplay that blurs the old distinction between nature and nurture.

Growing apart

Epigenetic changes begin in the womb. Unique epigenetic profiles have been seen in identical twins born as early as 32 weeks. Once these tiny variations are in place, they can be amplified by experience.

Just how important they are is not clear. The ideal study would be to raise a batch of genetically identical clones in the same environment and see how they turn out. This clearly cannot be done with people, but it can be done with mice. In one such experiment, 40 mice spent three months living together in the same five-storey cage. At first the mice behaved in a similar way, but over time they began to diverge. The study did not look at the cause of these subtle differences but epigenetics is a prime candidate.

So there is more to who we are than our genes and environment. If you reset the clock to the moment you were conceived and reran your life over and over, you would turn out differently every time despite having the same genes and environment. Your mum and dad have a lot to answer for. But you can't pin everything on them.

Give me space

Personal space is bizarre. Everywhere we go, we have a bubble around us that carries an invisible 'keep off' sign. We know the sign is there because everyone else stays out of our space, and in return we don't invade theirs

Personal space shrinks when we're home and expands in unfamiliar places

Women prefer a greater distance when talking to men than the men do

Invasion of personal space triggers an adrenaline surge that boosts heart rate

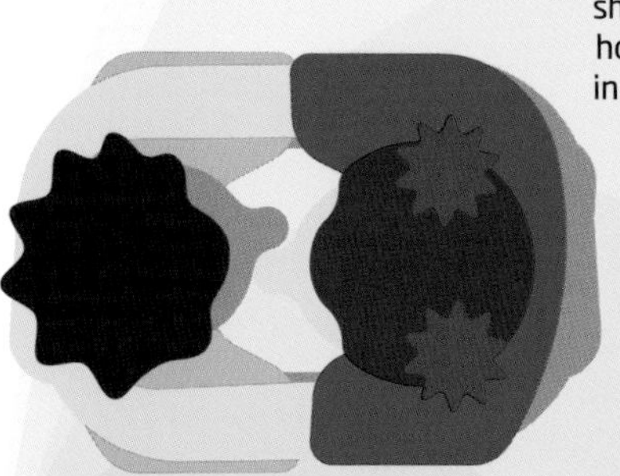

Close intimate distance

Up to 15 cm

This zone is for physical contact-nursing, love-making or fighting

Far intimate space

Up to 46 cm

Reserved for our closest relatives and friends.

Personal space

46 to 122 cm

This zone is for friends, close colleagues and partygoers

Some near-strangers can enter our intimate space, including dentists, doctors and hairdressers

Children and teenagers with autism are more likely to invade others' personal space

Women most dislike their space being invaded from the side, men from the front

People shrink their space for children and increase it for adults

When two people with different-sized personal spaces meet, one may complain of coldness, the other of feeling crowded

North Americans and Northern Europeans like large interpersonal distances. South Europeans stand closer, Arabs and Latin Americans closer still

In Westernised nations, two men prefer greater social distance than two women.

Social distance

122 to 370 cm

For impersonal and business conversations. Closer acquaintances may stand closer

Public distance

370 to 760 cm

A space reserved for talks or lectures to larger audiences

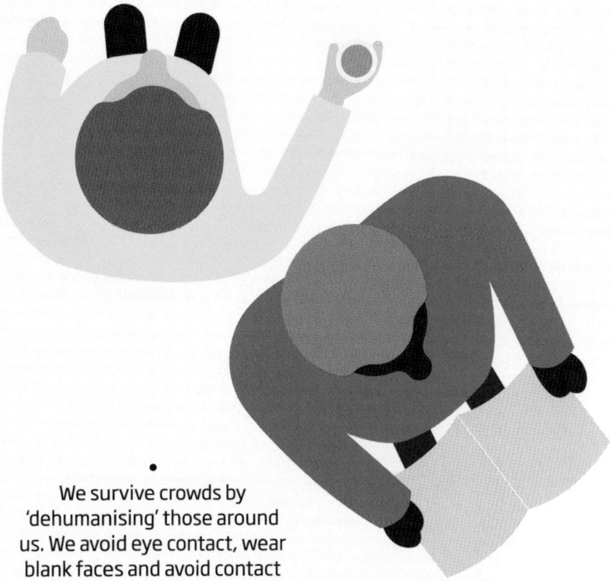

We survive crowds by 'dehumanising' those around us. We avoid eye contact, wear blank faces and avoid contact

Why you're good at everything . . .

How's your driving? If you are anything like the average person, you probably think it is pretty darn good. Surveys regularly find that about three-quarters of people believe themselves to be better than average behind the wheel.

That is highly unlikely to be true. Unless there are a handful of truly dreadful drivers out there, only about 50 per cent of us can be better than average. And yet if you ask people to rate themselves on almost any positive trait – competence, intelligence, honesty, originality, friendliness, reliability and many others – most put themselves in the better-than-average category. Ask them similar questions about negative traits and they will rate themselves as less likely than average to possess them.

Inflated opinion

This egotistic illusion is called the 'better-than-average effect'. It is extremely pervasive and highly resistant to contradictory evidence. Even people who are recovering in hospital after a self-inflicted car accident have described themselves as good drivers. The illusion also goes largely unnoticed. In an ironic twist, most people believe themselves to be more resistant than average to having an inflated opinion of themselves.

This is just one of a number of positive illusions that appear to be a common part of human nature. Another is unrealistic expectations about the future. Most people expect to live longer, healthier and more successful lives than average while underestimating their chances of getting divorced, falling ill or having an accident. And the more (or less) desirable the outcome, the stronger people believe it will (or won't) happen to them.

These unrealistic beliefs start in childhood thanks to our parents fawning over us, and don't stop once we grow up. Throughout life, we have an innate tendency to divide the world into 'us' and 'them'. As soon as you forge a connection with someone, you become part of their in-group – and humans are hard-wired to see members of their in-group more positively than they see others. In this way we all sign up to various mutual appreciation societies that exaggerate our virtues, ignore our faults and look down on outsiders. No wonder most of us feel excessively positive about ourselves.

But however deluded you are about yourself, chances are you are even more so about how you think others perceive you.

Everybody wonders and worries about how they come across to others, and most of us think we have a pretty good handle on it. But we don't. If you think of yourself as generous, for example, other people probably do too – just not quite to the extent you might like.

Deluded belief

From moment to moment, we are surprisingly poor at intuiting how we are coming across. This is largely down to something called the 'spotlight effect' – the deluded belief that everything we do and say is being closely observed and scrutinised by others. As a result, we blow everything out of proportion.

If you spill water on yourself so it looks like you peed yourself, you expect everyone to notice. But they don't because – and this might come as a shock – the world really doesn't revolve around you. People also assume that their emotional states are broadcast to all and sundry when in fact they are largely invisible.

This also works the other way. If you do or say something you think is especially witty or insightful, you are likely to overestimate the extent to which other people will notice.

The central problem is that you know yourself too well. You notice all kinds of subtle things about yourself that others simply don't.

Being immersed 24/7 in your own feelings and thoughts may give you greater insight into your private attributes, but it turns out to be a hindrance when it comes to rating aspects of your personality that are easily observed. If you ask somebody how considerate they typically are, they're likely to focus on how considerate they intend to be instead of how considerate they actually are. Other people simply judge us on what they see, making their assessments more accurate than our own.

This is compounded by the fact that we have terrible difficulty guessing what other people are thinking. Lacking the ability to read other people's minds, we fall back on reading the runes of their faces and behaviour. But behaviour doesn't always reveal what people are thinking.

She's talking already

The delusion that we are above average at almost everything isn't something we reserve for ourselves. We also inflate our opinions of loved ones. Around 95 per cent of people rate their partner as smarter, more attractive, warmer and funnier than average. And as anyone who has endured a dinner party of thirty-somethings will testify, parents of small children almost universally rate theirs as cleverer, cuter and more developmentally advanced than their peers. Which is tedious, but probably a good job from an evolutionary perspective.

The illusion of insight

Surprisingly, our lack of insight doesn't disappear when we're around people we know well: accuracy does go up, but only slightly. There is even evidence that our ability to read the mind of a spouse actually drops after the first year of marriage. Familiarity can create the illusion of insight. People are often better at knowing how well they're communicating with a stranger.

Perhaps the area where we have the least self-knowledge is physical attractiveness. Everybody knows what they look like, but when it comes to judging how we look, we're absolutely hopeless. Ask people to locate a photograph of themselves in a sea of faces and they find it faster if their own phizog has been altered to make it look more attractive, suggesting that we all think we're better-looking than we actually are. What's more, when psychologists ask people to score how attractively they will be rated by others, and check it against the actual ratings, they find almost no correlation.

Far from being pathological, though, positive illusions are viewed as being a marker of a healthy mind. The only people who appear immune are those with clinical depression, a state known as 'depressive realism'. Whether they are realists because they are depressed or depressed because they are realists is not clear.

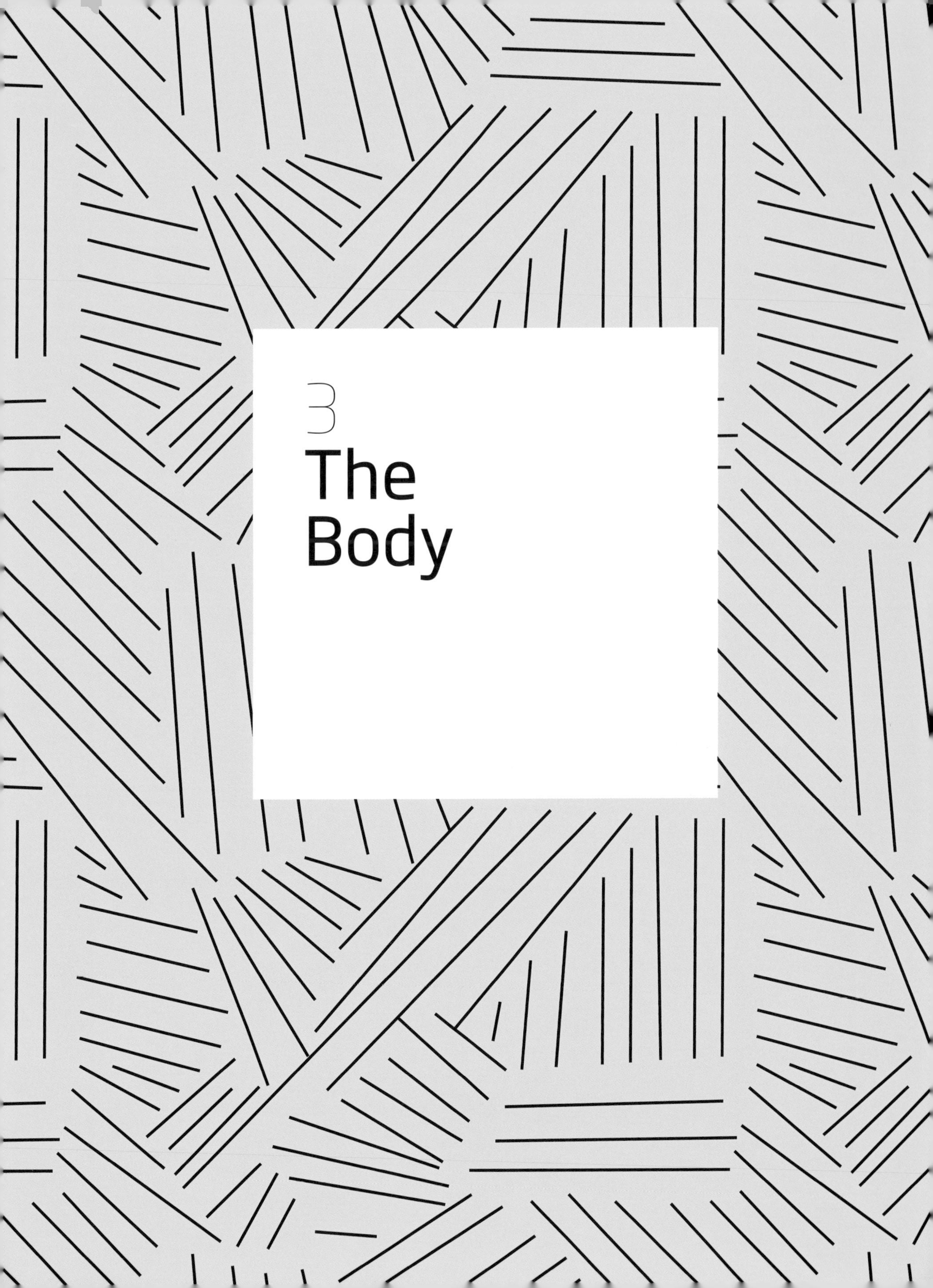

3
The Body

The human dimension

One head, two arms, two legs, ten fingers and toes, one heart slightly to the left of the chest and a kidney on either side ... The bodies of most human beings conform to a standard blueprint. Yet one look at the members of a crowd will tell you there's plenty of latitude within that pattern. Physically, humans are an astonishingly variable species. Only dogs are more so, and that is because we bred them that way. Humans, by contrast, have relied on natural selection and environment to make them the way they are.

No attribute differs more obviously than height. The shortest people in the world are the Mbuti people of the Democratic Republic of the Congo; an average man is just 1.37 metres (4 feet 6) tall. In comparison, men born in the Netherlands in the last quarter of the twentieth century average 1.83 metres (6 foot) – the tallest in the world.

On average, for people of European descent, women are 1.65 metres in height and men 1.78 metres. Height is one of the most heritable of all human traits – in other words, it's in our genes. But this is overlaid with environmental impacts.

Tall and cool

The evolutionary story of stature can be traced way back. Our ancestors 1.9 million years ago were even taller than the average Dutchman, with long legs and narrow bodies. These adaptations are thought to have arisen to help them keep cool as they walked and ran long distances to find food. Humans rely largely on sweating to cool down, and in hot dry climates a large ratio of surface area to volume increases heat loss.

As humans migrated towards the poles, however, they became shorter and stockier, with broad ribcages and pelvises. Again this makes thermodynamic sense: a shorter, wider body has a lower ratio of surface area to volume, so retains more heat.

The Mbuti people also owe their stature to heat regulation. Sweating is not very effective in humid environments where there is little or no air movement. The best way to prevent overheating is to limit the amount of heat you produce in the first place, by staying small.

Such general patterns linking stature to climate still hold, but a better predictor of height comes from looking at your family. Genetics appears to account for 80 per cent of the variation in height

Same but different

As well as the normal variation in our bodies, physical appearance can vary in astonishing ways. About 6 in 100 babies (mostly boys) are born with an extra nipple while 1 in 500 have a sixth digit on one or both hands. A very small number of people are born with their internal organs the wrong way round - heart and spleen on the right, liver on the left, and so on. Most are totally healthy but usually wear a medical tag in case they need emergency surgery. And if you thought every individual has unique fingerprints, think again. A rare genetic condition called adermatoglyphia leaves some people with no fingerprints at all.

within a population. The other 20 per cent is down to environment, especially health and food. Illness and malnutrition in a pregnant woman, can inhibit the growth of her fetus. Infections and lack of nutrients can impair the growth of children and adolescents.

Height limit

That 20 per cent largely accounts for why humans as a species have been getting taller. The average Dutchman, for instance, was 16 centimetres taller in 1990 than in 1860 thanks to better nutrition and healthcare. The same factors are also behind the two biggest gains in height seen in the past 100 years: South Korean women grew an average 20 centimetres taller while Iranian men added an extra 16.5 centimetres. Among well-nourished Westerners, however, the rate of growth in stature has been slowing for decades, suggesting that there is a limit to how tall our genes will let us grow.

If height is variable then body shape and weight are even more so. One estimate suggests that weight varies around the world by up to 50 per cent, even if the shortest pygmy populations are excluded. But globally, we're all moving in the same direction: outwards. Even in some of the poorest countries there have been increases in average body mass index.

Measures of BMI may even be underestimating our expanding waistlines. While children in the 1990s were about the same weight for their height as in the 1970s, they actually had 23 per cent more fat. The discrepancy was due in part to a reduction in the children's muscle mass.

One of the most curious findings from research into bodies is the stark difference between our own trajectory towards flabbiness and what we think of as ideal bodies. Over the past 60 years, ideals have got thinner for women and more muscular for men. Think of box-office stars: Marilyn Monroe versus Jennifer Lawrence, or Humphrey Bogart versus Matt Damon. Evidence for these trends comes from the shape of male and female bodies in soft porn magazines; the assumption being that such magazines act as a barometer of what is thought to be ideal at the time.

Between 1953 and 1980, the vital statistics of *Playboy* centrefolds changed dramatically. Bust and hip sizes decreased, while waists increased. The BMI of centrefolds also dropped compared with the average for the general population. Male centrefolds in *Playgirl* also grew leaner and more muscular.

Exaggerated dolls

Even children's toys are not excluded. In the 1990s, researchers scaled up Barbie dolls to life size to highlight how unrealistic they were. To be a real-life Barbie, an ordinary woman would have to grow 50 centimetres taller, add 13 centimetres to her bust and lose 15 centimetres from her waist.

Likewise, G. I. Joe has the dimensions and definition of a bodybuilder. Even dolls modelled on real people haven't escaped. In 1978, the *Star Wars* figures of Luke Skywalker and Han Solo resembled the bodies of actors Mark Hamill and Harrison Ford. By 1997, they had become like bodybuilders. Princess Leia's breast size tripled.

As the gap between real and ideal grows ever wider, so has the popularity of quick fixes. Every year around 20 million people – 90 per cent of them women – choose cosmetic surgery. From breasts to bums, there is hardly a piece of anatomy that can't be enlarged, reduced or reshaped. Ken and Barbie, here we come.

Why is your physique unique?

Look at the people around you and you cannot fail to notice how physically different they all are. Faces, bodies, mannerisms: all appear to be unique.

Now consider the whole of humanity. There are about 7 billion of us alive now and about 100 billion people have lived and died in the past 50,000 years. As far as we know each of them is, or was, a total one-off. The same applies to all those yet to be born.

That is a staggering amount of variation. But as we delve deeper into our biology and search for ever more sophisticated ways to verify people's identity, the ways in which we are all unique are being uncovered. So your mother was right: you are very special indeed. But don't just take her word for it.

Vast variation

The obvious place to start is DNA. And it is true: DNA does make you unique – up to a point. To get a measure of just how different you are genetically from everybody else try these numbers for size.

Our species is genetically very uniform. We all have about 99.5 per cent of our DNA in common, leaving just 0.5 per cent to account for all our myriad differences. Is it enough to account for the variation we see?

In theory, yes. The human genome contains approximately 3.2 billion letters of the DNA code; 0.5 per cent of that is about 16 million letters. The code has four letters, so the number of possible combinations is four raised to the power of 16 million – an absolutely vast number of possible genomes, more than enough to go around the world's inhabitants many times over. The chance of anyone having exactly the same genome as you is as good as zero.

That is even true of identical twins. Although they are 100 per cent genetically identical at the time of conception, from that moment on their genomes diverge, and the older they get the more individual they become.

These differences come from slight changes and chance mutations every time DNA is copied. What isn't clear is the proportion of these genetic variations that actually make you different. Many occur in regions that don't make proteins or regulate gene expression. And even if they are in these important regions, many are likely to be neutral in their impact, altering neither a gene nor how it is expressed.

We do know, however, that tiny genetic differences can have large effects on physical traits such as eye colour. So it is safe to say that your uniqueness as a person starts with the genome. But it is far from being the whole story. A case in point is fingerprints.

Everyone knows that fingerprints are unique, so it might come as no surprise that their size and shape is largely determined by genes. But the developing fetus's fingerprints are also altered by subtle factors like the pressure of the walls of the womb and even the sloshing of amniotic fluid.

This means that while the fingerprints of identical twins can be very similar there are enough differences to tell them apart. The same goes for toe prints.

An ear in a million

The same is also true of ears. You probably haven't paid that much attention to the shape of yours, but if you look in the mirror you will see that one is very slightly different from the other. Not only that, but each of your ears is different from everyone else's.

This, too, is a combination of genes and environment. Genes map out the general shape, while the environment in the uterus, such as how

The other you

One aspect of your uniqueness isn't, strictly speaking, part of you at all. It comes from the 100 trillion bacteria that live both on and in you. They massively outnumber the body's cells and in genetic terms they are even more dominant: microbes account for 3.3 million genes, compared with your measly 23,000. In other words, you're only 0.7 per cent human.

Of the more than 1,000 species that commonly live in and on the human body, each of us harbours only 150 or so, mostly in the gut. And everyone's bacterial population is made up of a different cast of characters.

Skin bacteria, too, vary from person to person. Even identical twins, who are difficult to distinguish on the basis of DNA, are easy to tell apart when you check out their bacterial companions.

the fetus lies, supplies the finishing touches. Once formed, ears hardly change shape as we (and they) grow and age.

Eyes are similarly unique. But as anyone who has their dad's eyes knows, the appearance of the iris runs in families. So how can eyes that appear to be the spitting image of the rest of the family's be totally unique?

The answer lies in the complexity of the iris's structure, a tangled mesh of muscles, ligaments, blood vessels and pigment cells, which give it colour, depth, furrows, ridges and spots, that forms at random during development. By this measure, each of your eyes is as different from the other as it is from anyone else's. The way you walk and talk are all your own too, determined largely by physical attributes such as the length of your legs and the shape of your larynx.

Your beating heart

There is another less visible way that everyone is unique. You wouldn't notice by putting your ear to somebody's chest, but no two hearts beat as one.

An electrocardiogram (ECG) records three peaks: the P wave, which is the impulse that contracts the upper chambers, the QRS complex, which is the stronger contraction of the lower chambers, and the much smaller T wave as the heart relaxes. Each heart varies in size and shape, so these features vary from person to person. And while the spacing of the peaks changes as the heart rate speeds up with exercise or stress, the individual signature can still be discerned.

All told, you're a physical specimen like no other. But one attribute is less distinct than you might think: your face. Though we find it easy to recognise people by face alone, the world is actually full of doppelgängers. One analysis of several thousand faces found that 92 per cent of them had at least one lookalike that both humans and facial recognition software struggled to tell apart. So even though you are definitely unique, your face is not.

What's in a face?

Faces are biological billboards that tell us obvious things about the age and sex of the owner and their emotional state. Less well known is that faces send out a host of hidden signals that we have evolved to read

KEY

Male and female faces | **Male faces** | **Female faces**

For all ethnicities, healthy faces tend to have a yellowish tint, red cheeks and lighter skin under the eyes. The yellow comes from carotenoids found in fruit and veg, which boost the immune system. Lighter skin is associated with the ability to absorb vitamin D and redness implies a good circulation and an active lifestyle

Baby-faced people – with large eyes, large heads and small jaws – are assumed to be warmer, more honest and naïve than mature-faced people. We don't know if these assumptions are correct but they matter: baby-faced defendants are more likely to win civil and criminal cases in court

When testosterone levels increase, men tend to be more attracted to feminine features in faces

Overlay lots of male or female faces and the resulting image tends to be perceived as more attractive than the constituent faces. For women,these faces have higher cheek bones, thinner jaw and larger eyes than average

Women who consider themselves to be particularly attractive show stronger preferences for masculine-looking men

Women's faces redden slightly around the time of ovulation. It seems to be caused by a rise in the sex hormone oestradiol, which may dilate blood vessels in the cheeks. This reddening may be one reason why men find women more attractive at times when they are most likely to conceive

Women tend to be more attracted to masculine faces around the time of ovulation. At other times, they prefer less masculine-looking 'caring and sharing' men who will make good long-term partners

Typical female features, such as a small nose and chin, large eyes and full lips, reflect high levels of oestrogen

Testosterone makes a face more masculine, with a prominent jaw and brow ridge. These features may also advertise a strong immune system because testosterone suppresses immunity and fit, masculine-looking men must compensate for this effect

We all tend to find symmetrical faces more appealing, but why? One notion is that it reflects the ability to grow straight despite life's challenges, including stress and infection. In effect, symmetry is saying 'I've got a great immune system!'

Activity in the brain's reward centres increases when a person smiles and looks directly at you, suggesting that you gauge attractiveness in part by how likely it is that the person wants to interact with you

Politicians who look competent are more likely to win elections than their more baby-faced opponents. Whether they really are more competent is still unclear

A glance at a person's face is enough for most people to judge whether a person would be a good bet for a committed relationship or a fling

We perceive people without much fat on their faces as more healthy and attractive. These people are less likely to suffer infections and have lower rates of depression and anxiety

People tend to credit individuals who have attractive faces with positive personality characteristics, such as higher intelligence. The single best predictor of satisfaction with a blind date is facial attractiveness

A trustworthy face tends to have a U-shaped mouth and eyes that form an almost surprised look. An untrustworthy face has a mouth that curls down at the corners and eyebrows that form a V. How trustworthy the owners of these faces really are is unclear

Masculine male faces signal dominance. More masculine-looking men reach higher ranks in the US Army

Men with very masculine faces are often seen as cold and dishonest. They are perceived as poor fathers and less likely to commit to a relationship. True to form, they are less likely to marry and more likely to divorce

Men and women rate full beards highest for parenting ability and healthiness

Women generally find men with clean-shaven faces, light stubble and heavy beards less attractive than those with heavy stubble

Men with high testosterone tend to have faces with a large width-to-height ratio. This ratio turns out to be a good predictor of aggressiveness

NOTE: MOST RESEARCH ON FACES HAS BEEN CONDUCTED ON HETEROSEXUAL MEN AND WOMEN WITH CONVENTIONAL GENDER IDENTITIES

What does your body language say about you?

'A body language expert's dream.' That is how one British newspaper described the Manchester United manager José Mourinho after a series of 'increasingly weird greetings' to rivals. His embrace of Tottenham's Mauricio Pochettino was 'part-affection, part control-battle'; his cursory hug with Jurgen Klopp of Liverpool the 'ultimate put-down' and his handshake with Arsenal's Arsène Wenger revealed a 'complete lack of empathy and rapport.'

Popular culture is full of such supposed insights about body language. But can we really read a person's thoughts and emotions from their body movements? Is there even such a thing as body language?

Disposing of the chaff

One statistic that is often trotted out by body-language experts is that 93 per cent of our communication is non-verbal. This figure comes from research done in the 1960s by psychologist Albert Mehrabian. He found that when there was a mismatch between words and deeds – for example, saying the word 'brute' in a positive tone and with a smile – people were more inclined to believe the non-verbal cues. From these experiments Mehrabian derived the '7%–38%–55% rule': 7 per cent of an emotional message comes from words, 38 per cent from the tone and the other 55 per cent from body language.

Mehrabian has spent much of the rest of his career fighting rampant misrepresentation of his finding. He says it only applies in very specific circumstances – when someone is talking about their likes and dislikes – and cringes every time he hears it applied to communication in general.

Mehrabian's rule isn't the only well-known fact about body language that turns out to be bunk. Another is that liars give themselves away with physical 'tells', such as looking to the right, fidgeting, holding their own hands or scratching their nose. Numerous studies have found that none of this is true. During appeals for information about missing persons, for example, people who later turn out to have been involved in the disappearance do not give themselves away in this way.

However, if you looked closely, you might find they are sending out some subtle bodily hints that they are lying – dilated pupils, fiddling with objects and scratching. The problem is, truth tellers also do these things, though they do them less often. They are not signs of lying but of general emotional discomfort. This is perhaps why most people are bad at spotting liars. Even professional lie detectors such as judges, police, forensic psychiatrists and FBI agents perform only marginally better than chance.

Many other well-known tells also don't reveal very much. Most people believe that crossed arms are a sign of defensiveness, which can be true, but it can also mean the opposite. Almost everyone judges a swaggering walk to signal an adventurous, extroverted, and trustworthy person while a slow, loose walk signals unflappability. However, there is no correlation between walks and these traits.

Victory pose

Nonetheless, there are some bits of body language that are reliable and universal signals. Athletes from all cultures strike the same pose in victory – arms aloft and chin raised – while losers hunch their shoulders. The same is true for athletes who have been blind from birth, suggesting that the poses are innate.

We can also glean accurate information about people from the way they move. Men judge a woman's walking and dancing as significantly sexier when she is in the most fertile part of her menstrual cycle. Women and heterosexual men, meanwhile, rate the dances of stronger men more highly than those of weaker men, which might be an adaptation for women to spot good mates and men to assess potential rivals.

Making the right impression

Ultimately though, it doesn't really matter what your body language actually reveals. What matters is what other people think it reveals. So if you want to come across as unflappable, adopt a slow and relaxed gait.

There are many other tricks that may help make the right impression. For example, people in job interviews who sit still, hold eye contact, smile and nod along with the conversation are more likely to be successful in winning the post. Those people whose gaze wanders or who avoid eye contact, keep their head still and don't change their expression much are more likely to be rejected. 'Mirroring' – subtly copying another person's gestures as you interact with them – is also a good way of building rapport.

Faked body language can even lead you to fool yourself. In one famous experiment, psychologists asked volunteers to hold either a 'high-power' or 'low-power' pose for 2 minutes. Afterwards, they played a gambling game where the odds of winning were 50:50. Those who had held high-power poses were significantly more likely to gamble. The researchers also took saliva samples to test the levels of testosterone and cortisol – the power and stress hormones, respectively. Power posers had elevated testosterone and decreased cortisol.

Moody movements

This is not the only way that body language can influence how you feel. Sitting up straight creates positive emotions, while hunched shoulders lead you to feel down. Faking a smile makes you feel happier, while frowning has the opposite effect. People who have Botox injections that prevent them from frowning feel generally happier. So even if Jose Mourinho's touchline antics don't reveal very much after all, when he cracks that smile, you know he's a happy manager.

Not flirting, but judging

Everybody recognises flirtatious body language, but using it to assess whether somebody fancies you can be risky. There is a popular notion that women signal interest in a man by touching their hair, tidying their clothes, nodding and making eye contact. That is true – but they also make the signals after meeting a man they don't fancy at all. Such flirting is only a sign of real interest if it keeps going for more than 4 minutes. Women unwittingly use this body language to keep a man talking until they can work out whether he is worth getting to know.

Love your extremities

No parts of our anatomy are perhaps so underrated as our hands and feet. Both are precision instruments that are essential to our success as a species. Our hands are tools for actively exploring and manipulating the world with both dexterity and strength. Our feet are beautifully adapted to upright walking and long-distance running.

Hold your hand out in front of you and you are looking at something uniquely human. Compared with even our closest evolutionary cousins we have much greater control of our fingers, and an amazing rotating thumb. Touch the pad of your thumb onto the pads of your other four fingers in turn and you are doing something that no other primate can manage. This is the famous 'opposable thumb', which allows us to hold and delicately manipulate almost any object. Chimps and gorillas also have opposable thumbs but do not have the range of movement that we do.

Our hands are also very strong, giving us the 'forceful precision grip' which lets us manipulate heavy and oddly shaped objects between thumb and fingers. Chimpanzee thumbs are too short to achieve such precision.

Our hands evolved quite quickly after we split from our last common ancestor with chimps. The presumed forebears of humans, the australopithecines who lived in Africa between 2 and 3 million years ago, did not have anatomically modern hands. Yet scans of their finger bones suggest they were capable of human-like grip.

Evolution probably favoured this grip to let our ancestors make and use tools. The earliest found so far are crude jagged stones made some 3.3 million years ago, which could have been used for digging, scraping or cutting.

Coordinated control

Another necessity for dexterity was brain power. Fashioning tools takes not only the right tendons, muscles and bones but also the ability to control them and to process feedback from the eyes and touch sensors in the hand.

Arch enemies

Not everybody is endowed with athlete's feet. Around 1 in 13 people are flat-footed: their midfoot comes into contact with the ground as they walk. This is surprising because a raised arch has been seen as an important adaptation for upright walking. As our ancestors evolved bipedalism, their feet developed longitudinal and transverse arches which created flexible rigidity, helping them propel themselves forward as they lifted their heel and pushed on the ball of the foot. Flat feet used to be considered a handicap – grounds to be excluded from military service, for example. We now know this was wrong: flat feet do not always cause pain or poor function.

Tool-making required our ancestors to realise that modifying one stone with another could make an even better tool. That's some conceptual leap. So the tools our ancestors made give us insight not only into their hands but also their mental abilities. Those ancient artisans would also have had to use their hands independently for different tasks – one to support a stone and the other to strike it. This is a skill that chimpanzees struggle with.

The interdependence of hand, eye and brain is still evident today. With 17,000 sensory receptors of various types, each hand has a capacity similar to that of the eye. Women, by virtue of their smaller hands, generally have greater touch sensitivity than men. An unduly large amount of the brain's sensory cortex, which receives and interprets sensory signals from around the body, is devoted to the hand. The same is true of the motor cortex, which plans and controls voluntary movements.

The ability to use our hands and coordinate their movements is not instinctive but has to be learned. Newborn babies have a delightful habit of grasping your finger with their whole hand. This is all they can manage until about 12 months when they learn to grasp with finger and thumb. Fine control does not develop fully until at least 10 years.

Once we have full control, possibilities snowball, whether it's playing a piano, making a chair or communicating with others. Hand gestures may have been a forerunner of spoken language and they are a lifeline for millions of deaf people.

Plates of meat

If we know lots about the human hand, the same cannot be said of the foot. Its basic anatomy is no mystery: it contains 26 bones and more than 100 muscles, tendons and ligaments. But how these parts work together is still an unfinished picture. Until recently, the few studies on feet were carried out on Westerners, who mostly wear shoes. But it turns out that shoes are deeply distorting: compared with unshod feet, Western feet are outliers, deformed by being squeezed out of shape.

Lifelong barefoot walkers, then, are ideal research subjects for anatomists to study. They tend to have relatively wide feet and pressure is distributed fairly evenly across the parts that touch the ground when walking. The obvious assumption is that these are adaptations for upright walking, which evolved soon after our lineage split from chimps. They also enable something that we humans are awesomely adept at: running.

At first glance the idea that we excel at running sounds unlikely. Usain Bolt can briefly hit a maximum velocity of about 45 kilometres per hour. Cheetahs can easily double that; greyhounds, horses and even chimps can beat it too. Mo Farah won the 2012 Olympic 10,000 metres in just over 27 minutes. A racehorse could run the same distance in less than 20 minutes.

Beyond 10 kilometres, though, the playing field starts to level out. At marathon distances and beyond humans are up there with the best. A well-conditioned athlete can run at 15 kilometres per hour for several hours, which is comparable to nature's endurance specialists, including wild dogs, zebras, antelopes and wildebeest.

This ability depends on anatomical adaptations to the feet, legs, hips, spine and rib cage that appeared in our lineage about 2 million years ago. The conclusion is that the human body is specialised for endurance running, perhaps as an adaptation for hunting (by running prey to exhaustion) or scavenging (allowing us to compete with dogs and hyenas for widely dispersed carcasses). You may not think so but, baby, you were born to run.

How old are your eyes, ears and blood . . .?

Imagine you buy a bike. After a while, the brake pads wear out and you replace them. Not long after, you get new tyres. The chain breaks, somebody nicks the back wheel, the handlebars snap, the seat wears out and the front wheel rusts away. All of these you replace, and more, until the only thing left of the original bike is the frame. Is it still the same bike?

Maybe it doesn't matter. But imagine that the same applied to your muscles, skin, guts, bones, heart and brain. If these are not the ones you were born with, are you the same person?

Replaceable body parts

Wear and tear is a fact of life. According to folklore your entire body is replaced every seven years. That isn't quite right, but there is little doubt that many cells are constantly being replaced. So do you really renew your entire body every few years? If so, how many bodies do you go through in a lifetime? If you live to a ripe old age, is there anything left of the original 'you'?

Answering those questions is surprisingly difficult. For a long time, the best guesses came from experiments feeding radioactive nucleotides, the building blocks of DNA, to rats and mice. New cells incorporated the chemicals into their DNA, allowing researchers to calculate the rate of cell turnover.

These experiments revealed that rats and mice routinely renew their body parts. But what about humans? Feeding them radioactive chemicals and then killing them is not an option. There are indirect methods for some tissues, such as seeing how long transfused blood cells last. But until relatively recently there was no way to directly age a cell in a human body.

That has now changed, thanks entirely to an ingenious method similar to carbon dating – with a cold war twist.

Carbon dating relies on measuring the amount of carbon-14 in organic matter. Carbon-14, a rare and weakly radioactive isotope of carbon, is continually produced in the atmosphere when cosmic rays smash into nitrogen nuclei. It eventually decays back to nitrogen, with a half-life of 5,730 years.

But before it decays, it can be taken up by plants and converted into sugars. Animals eat the plants, and in this way all living things contain small amounts of carbon-14 – about one in a trillion carbon atoms in your body are carbon-14 rather than the normal carbon-12. At death, however, the organism stops taking in carbon-14, and what it already contains eventually decays away.

That slow decay is what makes carbon dating of archaeological samples possible. By measuring the ratio of carbon-14 to carbon-12 in something that was once alive you can estimate when it died – up to 60,000 years ago. For anything that died before this, levels today would be too low to be useful.

Slow decay, however, also makes the method imprecise. An archaeological radiocarbon date is accurate only to between 30 and 100 years – fine for ancient Egyptian artefacts but useless for dating cells in a human body.

Unexpected help

But we can use carbon-14 in a different way thanks to a unique episode in recent history – the cold war. Between 1955 and 1963, above-ground nuclear weapons tests launched masses of carbon-14 into the atmosphere. At its peak in 1963, atmospheric levels of carbon-14 reached twice the normal background level. This 'bomb spike' was accurately recorded at locations all over the world.

Renewal versus ageing

If a large proportion of your body is younger than you are, why do we age? Why don't we retain a smooth complexion into old age, or run as fast as a teenager?

The answer lies with mitochondrial DNA, which accumulates mutations at a faster rate than DNA in the nucleus. As soon as you are born, your mitochondria start taking hits – and there is nothing much you can do about it. So while your cells may be only a third as old as you are, your mitochondria are the same age. In skin, for instance, mitochondrial mutations are thought to be responsible for the gradual loss in the quality of collagen, the skin's scaffolding, which is why skin sags and wrinkles.

The carbon-14 was also absorbed into people's bodies, including their DNA, which offers a way to tell the age of their cells.

Most biomolecules are in a state of flux, but DNA is very stable: when a cell is born it gets a set of chromosomes that stay with it throughout its life. The level of carbon-14 in a living cell's DNA is therefore directly proportional to the level in the atmosphere at the time it was born, minus a tiny amount lost to decay. Before 1955 that level was always roughly the same. But during the bomb spike, atmospheric levels changed. What that means is that you can take cells born between 1955 and about 1990, measure the carbon-14 in their DNA, and then obtain an estimate of their date of birth.

Using this technique scientists have finally estimated how quickly a human body replaces itself.

It's tough on the front line

Front-line cells endure the roughest life and last the briefest time – these include cells lining the gut (5 days), the skin's surface (2 weeks) and red blood cells (120 days).

Muscle cells have an average age of 15.1 years, and those making up the body of the gut 15.9. The entire skeleton is replaced every few years. Heart muscle, however, does not seem to be renewed at all. So the common knowledge that the body replaces itself every seven years is not too far wide of the mark.

The brain is a different story, however – which might be why you still feel like you even though your body has been cast off and replaced. Neurons from the visual cortex and olfactory bulb, for example, are the same age as the person they come from.

But not all of your brain has been with you for life. In a region called the hippocampus, which is closely associated with memory, around 700 neurons are replaced every day. That's an annual turnover of about 0.6 per cent. If you are 50 years old, around a third of your hippocampus is younger than you are.

All told, regardless of how long you have been alive, your body is, on average, only about 15 – though like a (t)rusty old bike frame, some parts have been with you from the start.

Why are humans so hairy?

Good God, you're a hairy beast. You may not think you are, and compared with, say, a chimpanzee, you appear distinctly bald. But in fact your entire body (with the exception of the palms of your hands and the soles of year feet) is covered in hair. All told you have about 5 million follicles, about the same as chimps and other primates.

But the similarities stop there. Human hair is decidedly strange. Most of our body hair is so wispy and short as to be almost invisible, though in some places it is coarse and curly. Our head hair is almost uniquely long and flamboyant. We are pretty much the only animal to have hair that grows continuously for many years, and also to suffer the indignity of going bald. No wonder our relationship with our hair is a tangled one.

A duet of hairs

Human hair comes in two basic types: terminal hair, which grows on the scalp, eyebrows and eyelashes, and vellus hair, which is found everywhere else. Beyond that, the main difference between hair types is how long they grow for before the follicle runs out of steam. This is what determines their length and thickness.

Hair follicles go through cycles of growth and dormancy. During the growth phase the hair grows continuously at about 0.4 millimetres per day, getting longer and thicker in the process. But at some point the hair-producing cells die off and growth stops. The hair falls out and the follicle goes dormant for around six months before sprouting new hair-producing cells and entering a new growth phase.

The length of the growth phase is controlled by hormones. Leg hairs grow for about 2 months, which is why they are short and fine. Armpit hairs make it to six months, but head hairs grow non-stop for six years or more. That means head hair can theoretically grow to almost a metre in length.

Ideas abound as to why evolution has endowed us with such a unique combination of hair types. The leading one is that when our bipedal ancestors moved out of the forests and onto the searing heat of the savannah, they needed to keep their bodies cool while also sheltering their big brains from the sun. Body hair receded, removing an unnecessary insulating layer and also allowing the skin to be cooled by sweating. Hair loss was also presumably aided by technological innovations such as clothes, fire and cave-dwelling, which lessened the importance of fur for keeping warm at night. Head hair, meanwhile, became thicker and more luxuriant, protecting our ancestors' brains from the midday sun, and also retaining heat in the cold.

Genetic evidence suggests that we became furless around 1.7 million years ago. Around this time our ancestor *Homo erectus* was living on the baking savannah, which supports the thermoregulation hypothesis.

Look! No parasites

But it is not the only possibility. We may also have lost body hair to improve our ability to identify others and to make communication easier, or to resist disease, since fur is a prime habitat for parasites. There's also sexual selection, which was Darwin's preferred explanation. For whatever reason, the least hairy of our ancestors were considered the most attractive and so produced more offspring. Hair-free skin may have been a billboard to advertise good health and hence attract mates – a sexual come-on saying, 'Look how unblemished and parasite-free I am.'

Sexual selection could also explain our head hair. Most people find a strong, healthy head of hair attractive. Indeed, precisely because our head hair needs so much care, it makes a perfect hoarding upon which to advertise social and sexual status. Grooming is time-consuming, and so having well-groomed hair shows you are resourceful and have good social contacts. If this is correct, then the main function of head hair is to be cut and styled for maximum impact. This might explain why lank, unkempt locks are the mark of the social pariah.

Prehistoric styling

Some of the most ancient figurines do have coiffed hair. The oldest known three-dimensional representation of a human, the 25,000-year-old ivory Venus of Brassempouy, has elegant shoulder-length hair. And hair products are nothing new either. The 2,300-year-old Clonycavan Man, discovered in a bog in County Meath, Ireland, was wearing hair gel made of plant oils and pine resin.

Hair may also have been used to signal group identity. Throughout the ages and across cultures, we have used hairstyles as a mark of membership: think Roundheads, Rastafarians and rockabillies.

Enough about the hair on your head. Pubic hair is possibly even more unusual. Most primates have finer hair around their genitals than on the rest of their body, but adult humans are the exact opposite.

There's no accepted explanation. One possibility is that since thicker hair coincides with regions where we have apocrine (scent) glands, it may serve to concentrate or waft odours that signal sexual maturity. Pubic hair may also protect the genitals during sex and at other times – reducing chafing while walking, for example – and also helps keep our most sensitive regions warm and free of draughts.

Whatever it evolved for, many people now subject the hair in their pubic regions to as much grooming as the hair on their heads, while ruthlessly removing it from the rest of their bodies. Hairless we ain't – but not for lack of trying.

Baldylocks

Pity the human male: along with the stump-tailed macaque, he is the only primate to routinely suffer the indignity of a receding hairline.

By age thirty, a quarter of men have started to go bald, and by forty-five, half have. Male-pattern baldness has a predictable trajectory: hair begins to disappear on the temples, then on the top of the head, before staging a general retreat that leaves the entire pate hairless.

Except that it doesn't. Baldness is not about having no hair; it is about having the wrong sort of hair. Bald heads have just as many follicles as any other – about 100,000 – but the follicles have ceased to function properly and only produce colourless and wispy hairs.

Does your voice reveal secrets about you?

The year was 1927. The BBC was only five years old and radio was still a novelty. With disembodied voices floating across the airwaves into almost every living room in the country, British psychologist Tom Hatherley Pear wanted to discover what was going on in people's minds as they listened.

He recruited nine people – including his eleven-year-old daughter, a judge and a minister – to read a passage from *The Pickwick Papers*. The readings were broadcast across the UK on three consecutive nights; listeners were asked to cut out a form from the *Radio Times*, fill it out describing their impressions of the speakers, and send it in.

Nearly 5,000 people responded, many with detailed descriptions and even life stories of the speakers. Whether they were right or wrong, the strength of the impressions they formed was striking. Pear had identified something that we perhaps all know instinctively, that the voice can be powerfully suggestive. Whether you are eavesdropping from another room or taking a phone call at work, the way someone speaks can paint a clear picture of them in your mind's eye.

Highs and lows of attraction

Perhaps the most obvious information a voice can convey is the speaker's gender – chiefly through its pitch. In many species, males have longer vocal tracts than females, enabling them to produce deeper sounds, which seems to have the effect of making them appear bigger to potential rivals and mates. This may well be sexual selection in action: as females over time have chosen mates with more sonorous voices, so evolution has favoured longer vocal tracts in males.

In humans, men speak on average an octave lower than women. Studies show that women tend to find men with deeper voices more attractive, while men prefer higher voices in women. The pitch of a woman's voice even increases slightly two days before ovulation, possibly enhancing her allure.

Pitch aside, other subtle signals of gender may be at work. For example, two voices speaking at the same frequency can sound either masculine or feminine depending on the way they pronounce the sibilant 's' sound at the end of words such as 'centuries'.

Another characteristic that can be gleaned from a voice is the speaker's age. As we get older, we speak more slowly and, as muscle tone decreases,

Lost for words

If our voices are our 'auditory faces', then it is no wonder that speech impediments can have a devastating effect on people's lives. One of the most common is stammering or stuttering, which affects around 70 million people worldwide. Every language has a word for it, and stammerers are often discriminated against and subject to misguided attempts to solve the problem. Despite this, its cause is largely unknown. Recent research suggests it may have its basis in subtle brain differences, which may eventually lead to a much-needed cure.

our speech can get weaker and breathier. Some people opt for 'voice lift' surgery to make them sound young again. But these cues are not iron-clad indicators of age.

One of the more intriguing aspects of the way we speak is that people often believe they can describe a speaker's physical characteristics from their voice alone. It's true that certain physical attributes influence the volume, pitch and timbre of the voice – the shape of the lips, jaw, nose and chest, for example. But reading these from a voice is very difficult. If asked to pair voices and faces, people perform only slightly better than chance. Curiously, however, we can guess people's height based on their voice.

The sound of leadership

Things get really thorny when it comes to intuiting people's psychological features. These judgements seem to be based on crude stereotypes.

For both sexes, a deeper voice is widely assumed to mean greater competence and leadership abilities. Among male chief executives of US companies, for example, those with deeper voices tend to run bigger firms and earn over $150,000 a year more than men with higher-pitched voices. Women's voices are judged similarly. People asked to rate female politicians by voice alone consistently preferred – and said they would vote for – those with deeper voices. Advisers to the late UK Prime Minister Margaret Thatcher were apparently aware of this effect: they coached her to deepen her voice to increase her authority.

But going deep can go too far. Many people (perhaps unwittingly) employ 'vocal fry' to enhance the authority of their voice. This involves dropping to a gravelly pitch. But it does not go down well with listeners of either sex. On attributes such as attractiveness, education, competence and trustworthiness, vocal fry is perceived negatively relative to a normal voice.

Accent bias

An even more egregious influence on our judgements is accent. Our reliance on accents probably had an evolutionary benefit by helping us to create our cultural identities, so we could identify our in-group from out-groups. Today the inferences we make from accents are complex, influencing our perceptions of attractiveness, prestige and intelligence. These can leave us wide open to prejudice.

Stereotypical accents have long been used in TV and film to lead the audience. Think of the grating New York twang of Janice Goralnik, Chandler Bing's on-off girlfriend in *Friends*, or the 'Worzel Gummidge' accents often used to signify stupidity. This may be acceptable for entertainment, but in the real world the same biases can have serious consequences. Play British people recordings of an accused man and they are more likely to find him guilty if he speaks with a Birmingham accent rather than the Queen's English. And woe betide anyone with a thick accent that makes them harder to understand: listeners tend to mistrust what they say.

Our voices have an important impact on our lives. Not for nothing have they been described as our 'auditory faces'. Of course, if you don't like your face, there's not a lot you can do about it – at least easily. But voices are different. Many of us subtly and often unwittingly change how we speak according to social circumstances. So if you want to pimp your auditory face, learn to speak proper.

Why you may not be alone in your body

You probably think you will always be in your mother's heart, and she in yours. And you'd be right – quite literally. After you were born, you probably left tiny bits of yourself inside your mother. And you got stuff from her, too: her cells take up residence in most of your organs, perhaps even your brain. They live there for years, decades even, meddling with your biology and your health. The same is true of your own children and your brothers and sisters.

Sure, your blood, skin, brain and lungs are made up of your own cells, but not entirely. Most of us are walking, talking patchworks of cells, with emissaries from our mother, children or even our siblings infiltrating every part of our bodies. Welcome to the bizarre world of microchimerism.

Mixing blood

The idea first emerged in the 1970s. It used to be thought that the blood of a pregnant woman and her fetus were kept completely separate. But then cells with the male Y chromosome were found in the blood of women who were carrying boys. The placenta, the organ that conveys nutrients, oxygen and waste products to and from the fetus, was also letting a few blood cells through. Today we know that these cells can lodge in our bodies long after birth.

Finding these microchimeric cells is hit and miss, not only because there are relatively few of them, but they also appear to move around the body, their numbers rising and falling in different organs at different times. Some think that with enough testing we would find that most, if not all people will turn out to be microchimeric.

Some questions are being answered. For example, the foreign cells seem to take up residence everywhere. Women carry cells from their children in every organ tested so far, including the brain, and these cells can survive for 40 years. There are also hints of health effects: people who have more microchimeric cells tend to suffer more of certain autoimmune diseases, but are at lower risk from diseases such as thyroid and breast cancer. They may even live longer.

Accelerated healing

How these cells exert their influence in some conditions is also slowly coming to light. Peripartum cardiomyopathy is a disease in which a woman's heart becomes weak and enlarged during pregnancy. Half of these women recover, but we did not know why.

To find out more, researchers in the US created a mouse model of the disease. They mated female mice with males carrying a green fluorescent protein in their cells. Half their fetuses inherited the protein, making their cells easy to spot in the mothers. Then the team induced heart attacks in the pregnant mice. Astonishingly, the fetal cells homed in on the damaged heart tissue, where they accelerated healing by turning into heart muscle and blood-vessel cells.

The fetuses, it turns out, provide a reservoir of embryonic stem cells for the mothers. This raises the promise of new therapies: could fetal cells be used to treat other forms of heart disease?

Intriguingly, from other animal work we now know that fetal cells can turn into neurons in their mother's brains. Whether these cells are repairing damage or are part of normal development is still to be discovered. It also invites the question of whether maternal cells play an active role in the brains of their offspring.

The health impacts of microchimeric cells may even help in the birth of the next-but-one

Extreme mixing

In 1998, a fifty-two-year-old woman in Massachusetts made a macabre discovery about her past. She needed a kidney transplant and her three sons were tissue-typed to see if one of them could be a donor. Incredibly, the results showed that she was not their biological mother. It took two years to solve, but her doctors eventually discovered that she was a chimera, a mixture of non-identical twin sisters who fused in the womb and grew into a single body. Cells from one twin had come to dominate the woman's blood, which was used for tissue typing. But in her other tissues, including her ovaries, cells from both sisters lived amicably side by side. Full-blown chimerism is much rarer than microchimerism - only around 30 cases have been discovered. But there are probably many more out there who simply do not know that they are a mixture of two different people.

generation. Pre-eclampsia is a nasty condition suffered in about 6 per cent of pregnancies. It affects women in their third trimester and can lead to poor outcomes for both mother and baby. Scientists studying women at risk of pre-eclampsia found that those who developed the disorder carried no cells from their mothers. Yet nearly a third of women who avoided the disease were carriers. It remains to be seen what role – if any – the mothers' cells are playing in their pregnant daughters.

Looking for a reason

Beyond its health effects, other questions remain about microchimerism: why, for example, does it happen at all? There could be an evolutionary mechanism at play here: if by passing on a few cells to its mother a fetus can improve its chances of survival then evolution might well favour the process.

Then there's the question of how alien cells avoid the host's immune system which, after all, is designed to see off invaders. We know that microchimeric cells can turn into a type of immune cell. Could it be that they somehow embed themselves within our body's defences, becoming part of our 'biological self'? Understanding this mechanism is important because it could help to stop rejection during organ transplants.

Your biological self

Whatever the solutions to these puzzles, the notion of the biological self is badly in need of a rethink. We now know that our genome is stuffed with viral DNA (see 'The viruses that are now part of you', p 114) and our mouth, gut and skin are home to billions of microbes that live in relative harmony with us. Now we find our cells are not all our own. To misquote the poet John Donne, 'No man (or woman) is an island'. Today, it seems we are more like walking, talking ecosystems.

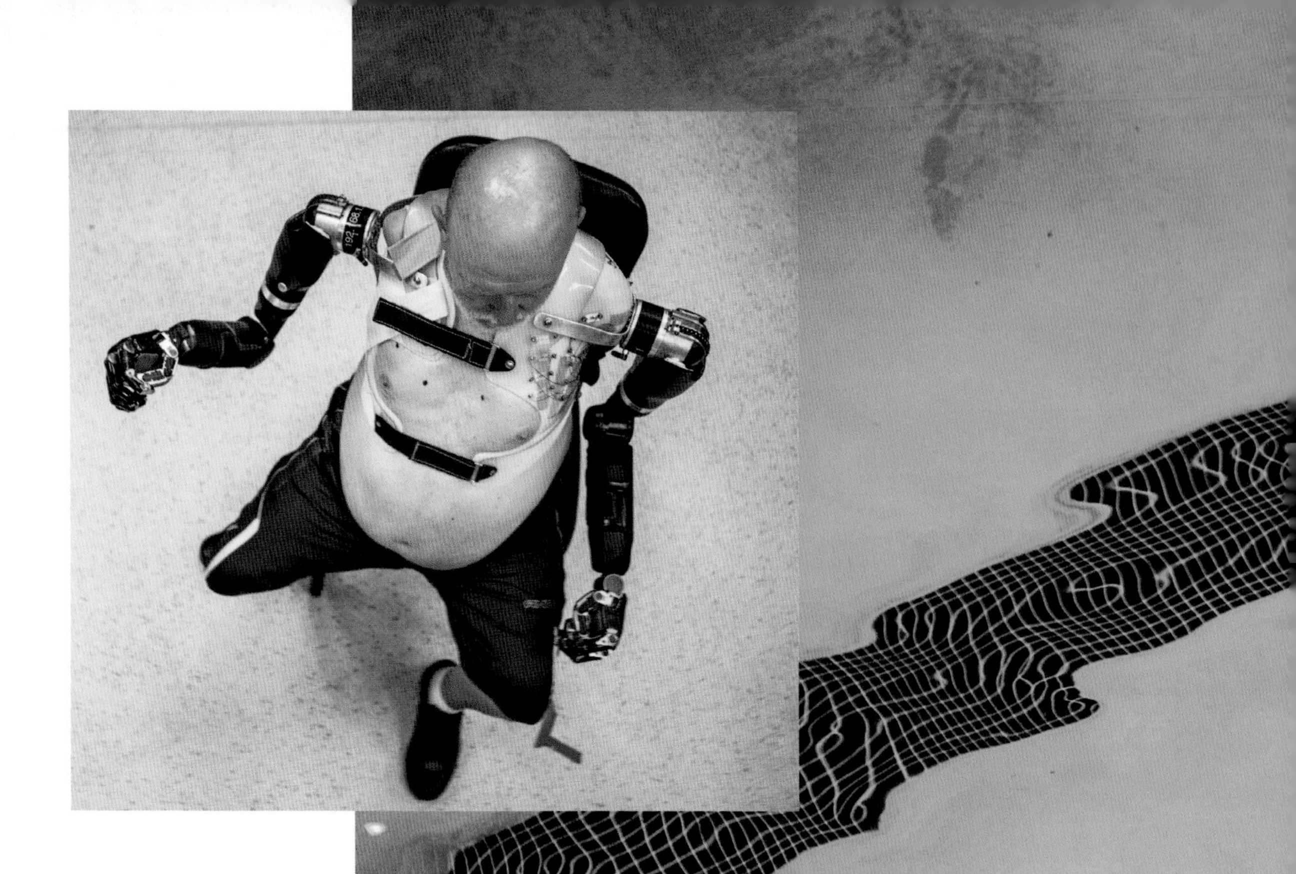

We can rebuild him
Learning to speak the bioelectric language of our nerves is starting to free us from the frailties of our bodies. Les Baugh, here seen swimming and sporting his prosthetic limbs, is a beneficiary of that knowledge. He lost both arms in a freak accident 40 years ago. In 2013 researchers from Johns Hopkins University in Baltimore took nerves that once made his arms move and reattached them to muscles in his chest. Today, those chest muscles act as amplifiers, boosting nerve signals from the brain which are then fed to a computer in his prosthetic limbs. Just by thinking about rotating his wrist or extending his elbow, Les can make his prosthetic arm obey his thoughts.

Credit: Zackary Canepari / Panos Pictures

Smart thinking needs a body

'I think therefore I am,' wrote René Descartes. He did not know the half of it.

Descartes's famous maxim was part of a treatise setting out the idea known as 'dualism' – that the mind and body are separate entities. This is a distinction that comes easily to us; we tend to view thinking as something done exclusively by the brain. Yet it is becoming increasingly clear that thinking involves the whole body. Without input from your body, your mind would be unable to generate a sense of self, process emotions or think high-minded thoughts about language and mathematics. Far from being just brawn, your body turns out to be the brains of the operation too.

The most basic aspect of the mind–body connection is the familiar yet puzzling sensation of embodiment. This is the feeling that we occupy and own our bodies, which is a central tenet of our sense of self and hence consciousness.

Embodiment is created by your brain as it processes sensory information, and it is surprisingly easy to manipulate. One particularly eerie experiment is the rubber-hand illusion. Place a rubber hand (or an inflated rubber glove) on a table in front of you, put your real hand out of sight, then ask somebody to stroke both rubber and real hands with identical motions. Watch the fake hand being stroked and you may start to feel that that it belongs to you. This is caused by a mismatch between visual and tactile sensations, which your brain resolves by going with what it can see.

A similar set-up can make people feel as though their entire body has been transferred to a dummy, or even a Barbie doll. And if somebody strokes your face in sync with a random face on a screen, the image can come to feel like your own reflection.

These illusions don't just tell us about how the brain and body combine to create embodiment, they also demonstrate how the body participates in mental processes – what is called 'embodied cognition'.

For example, people undergoing the Barbie doll illusion perceive their surroundings as much bigger than they really are, which suggests that embodiment affects how we interpret the sensory information hitting our eyes.

Embodied cognition also plays a role in emotions. You may think that you smile because you are happy, but in fact it works the other way round. Happy feelings arise in a large part from the physical sensation of smiling. Forcing a smile can change your mood, and people whose frown muscles have been frozen with Botox take longer to read sad or angry sentences than they did before receiving the treatment.

Moving marbles

Physical movements can also influence emotional processing. In one experiment, scientists asked people to move marbles from a box on a high shelf to one on a low shelf, or vice-versa, while talking about events that had positive or negative emotional significance – such as a time when they were proud or ashamed of themselves.

The volunteers were significantly faster at retrieving and retelling positive stories when they were moving marbles upwards, and vice versa. And when they were asked neutral questions – such as 'What happened yesterday?' – they were more likely to talk about positive happenings when they were moving marbles upwards, and narrate negative stories when moving marbles downwards.

Embodied cognition might also be involved in abstract thought processes. Mathematical thinking, for example, seems somehow to piggyback on our experience of movement and space. When people are asked to read out a list of random numbers, their eye movements mirror their thought processes. If the next number is larger than the previous one, they usually move their eyes upwards and to the right before saying it. If it is smaller, they move their eyes downwards and to the left. The bigger the difference between the numbers, the bigger the horizontal movement. The correlation is so strong that the next number can be predicted quite accurately from the eye movement alone.

This association between quantity and spatial movement may be learned from an early age. A child watching a glass of water being filled up, or building blocks being piled up, will learn that increasing height means greater quantity, leading to an intuition that up is more.

Creative posturing

Truman Capote once described himself as a 'horizontal author', saying 'I can't think unless I'm lying down'. He might have had a point. Stress is the enemy of creativity, and we feel more relaxed on our backs. People solve anagrams in about 10 per cent less time when lying down compared with standing.

There are other ways to get those creative juices flowing. Bend your right arm at the elbow so you resemble Auguste Rodin's statue *The Thinker*. People who strike this pose perform better on a creative-thinking task, in which they have to find innovative uses for an everyday object.

Lateral eye movements can help too. It is possible that this temporarily encourages communication between the right and left hemispheres of the brain, which boosts creativity.

Unconscious climbing

Other studies show that language is also embodied. Every time we hear a verb, the brain seems to simulate the actions associated with its meaning. When someone says the word 'climb', for example, it activates the same neural regions that trigger arm muscles.

Even abstract notions such as 'good' and 'bad' appear to tap into our sense of embodiment. In another experiment, volunteers were asked to make judgements about animal-like objects called Fribbles. The Fribbles were presented in pairs, one on the left and one on the right, each slightly different from the other. Between each pair was an instruction such as 'Circle the Fribble who looks more intelligent' or 'Circle the Fribble who looks less honest'.

Most of the volunteers showed a leftward or rightward bias and, more often than not, the preference matched their handedness. Left-handed people preferred Fribbles on the left, and vice versa. The researchers concluded that we tend to associate goodness and virtue with our dominant hand.

All told, it is hard to deny that thought is inherently physical as well as mental – so much so that artificial intelligence researchers now think that true AI is impossible without a body. We act, therefore we think.

Can we escape the metabolism prison?

We humans seem to be divided into whippets and walruses. Whippets eat what they like and stay lean and svelte. Walruses pile on the pounds just by looking at food. It seems so unfair.

What separates these groups is in large part down to metabolism, the sum total of chemical reactions that keeps our bodies ticking over. So what do we know about these differences in metabolism, and can we use that knowledge to make us walruses a little less ... walrusy?

Storing energy

One of metabolism's chief jobs is to generate energy. The fats, carbohydrates and proteins we consume end up in our cells, where they enter an intricate network of biochemical pathways. The end product is the energy we use to function. Any excess energy is stored in one of two forms: glycogen in liver and muscle and, when that store is full, fat.

The tempo of cell metabolism – our metabolic rate – is controlled by hormones released by the thyroid gland, which sits at the front of the neck. People with an overactive thyroid eat lots, get very hot and are as thin as rakes. In contrast, those with underactive thyroids eat less, become cold and clammy and gain weight. These disorders affect about 1 in 1,000 men, but are more common in women, with 1 in 100 suffering from overactivity and 15 in 1,000 with an underactive thyroid.

Establishing a person's resting metabolic rate is no easy matter. It means putting them into a small room called a metabolic chamber for, say, a day, during which time their consumption of oxygen is measured, together with their outputs of heat, carbon dioxide and nitrogen waste. The result is a profile of their energy consumption and expenditure.

Measured this way, it seems paradoxical that when resting an obese person will expend more energy than a lean person. But then you realise that larger people have more cells to maintain. And it's not only the number of cells that matter, it's also the type. A kilogram of fat cells, for example, burns a measly 4 calories a day. The same mass of muscle burns a modest 13 calories a day. These pale into insignificance next to heart and kidney cells, a kilogram of which will burn 440 calories a day. Obese people tend to have larger organs, which

Hormone tweaking

One way to influence weight gain may come from hormonal control. We now know of a group of hormones that suppress appetite. Leptin is the best-known, released by fat cells in response to eating. Another is protein YY, which is released in the lower gut, again in response to eating. In recent years, the spotlight has switched to oxyntomodulin, another gut hormone that seems to suppress appetite and boost metabolism. Controlling metabolism through such chemicals will present a challenge because of the complexity of the relationships between them, and the effects they have on the brain and other organs. Still, some researchers think hormonal control may yet deliver real benefits.

is another reason their energy expenditure is higher.

Figures like these show why it's so difficult to boost metabolic rate by building muscle. A person replacing a kilogram of fat with a kilogram of muscle would burn only 9 calories extra a day: hardly a recipe for becoming a whippet. The numbers also reveal why men, who tend to have relatively less fat and more muscle than women, need to eat more calories than women. Finally, they demonstrate why people need fewer calories as they get older: on average, a twenty-year-old man carries 5 kilograms more muscle than a sixty-year-old.

Resting metabolic rate can take us only so far. We all know that the more a person exercises the more energy they burn, which begs the question of why some lazy people are skinny. A possible answer is that these people are more active than they realise. Research on self-confessed couch potatoes found that lean individuals sat for 2.5 hours a day less than mildly obese people. This difference is equivalent to 350 calories a day which, over time, is more than enough to make a lean person fat.

Naturally prone

Activity, though, is not the only factor here. Evidence suggests that some people are naturally more prone than others to putting on weight. Overfeed identical twins for an extended period and those differences start to emerge. Within twin pairs, weight gain tends to be similar. But between different sets of twins it can vary by as much as threefold. Such results strongly suggest that genes are influencing our tendency to gain weight.

Which genes matter is still moot, but scientists have identified five traits that reliably predict greater weight gain: poor fitness; low muscle mass; low levels of testosterone – which boosts muscle growth; being less responsive to the appetite-suppressing hormone leptin; and burning less fat as fuel when digesting food and absorbing its nutrients.

Some people's bodies, then, cope better than others with excess food by burning more calories and not laying down fat. The differences between these obesity-resistant individuals and the obesity-prone is not just in their immediate response to food. Two days after overeating, obesity-resistant people report that they go off energy-rich foods, such as cake. Their brains are also less responsive to images of such foods than those of obesity-prone people. What's more, two or three days after overeating, obesity-prone individuals become more lethargic, while obesity-resistant people stay their usual active selves.

Hidden organ

One other influence on metabolism needs teasing out: the role of our gut flora. Research has shown that when mice are given gut microbes from fat animals, they too get fat. And transplanting gut microbes from thin animals can promote weight loss in the recipients. Just how the microbiota achieves this effect is unknown, but it is almost as if there's an extra 'organ' directing weight gain in our gut.

Metabolism and its link to weight gain is extremely complicated. No single factor is likely to provide a silver bullet, but understanding all the influences does give us walruses some levers with which to control our weight. Will we ever be able to transform ourselves into whippets? At least for now, that is a wish too far.

How can fat make you thin?

'Eat less, move more!' could be the rallying cry of our age. In many countries, and not just developed ones, the availability of calorie-laden food has turned obesity into a major health hazard. The agents of this menace are white fat cells which accumulate around our bums and tums, storing excess energy until it's needed. Unfortunately, for the many who live modern sedentary lifestyles that energy is never used.

It's paradoxical then that some scientists think the answer to this problem may also lie in fat cells. Not white fat, but a relative called brown fat, which has a propensity for burning energy. In fact, brown-fat cells produce 300 times more heat per gram than any other cells in the body.

Baby fat

It used to be thought that brown-fat cells existed only in babies, who need help regulating their temperature. Their number certainly diminishes with age, but it turns out that many adults still have deposits around their collarbones, shoulders and upper back. This has raised the intriguing idea of using brown fat's furnace-like ability to burn up our stored energy. The amount burned would not have to be huge: often obesity arises over time from eating just a few calories a day too many.

Brown-fat cells are brown because they contain abnormally large numbers of mitochondria, the tiny structures that provide cells with energy. These mitochondria are also different from normal: they contain a protein called thermogenin, or UCP1, whose sole purpose is to generate heat.

Brown fat is activated by exposure to cold, so the best way to turn it to our advantage would be to regularly cool ourselves down. Mammals initially keep up their temperature in such conditions by shivering. But research in rodents shows that after repeated bouts of cold, brown fat takes over temperature regulation, so shivering decreases but energy expenditure stays the same.

This effect can work in humans too – at least if we're prepared to submit to temperatures of 16 °C for six hours a day over an extended period. But that sort of torture makes sense only if it guarantees weight loss and, unfortunately, the evidence for that is mixed. Short bouts of cold can boost brown-fat activity and reduce body fat, but this does not prove cause and effect. Results so far suggest that brown-fat cells may play only a small role in reducing body fat while other mechanisms, including shivering, may be at play.

Even if brown fat is shown definitively to promote weight loss, many people are likely to baulk at having to endure doses of low temperatures. Fortunately, some chemical compounds seem to be able to mimic the effect of cold (see box, opposite). One such substance is mirabegron, a drug developed to calm down an overactive bladder. It also increases the activity of brown-fat cells. The resting metabolic rate of 12 volunteers who took the drug increased by 203 calories a day. Intriguingly, mirabegron also broke down white fat.

Boosting brown fat

One potential problem shared by all these approaches is that many people, especially adults, have only a tiny number of brown-fat cells. So one question being asked is, can we increase our brown fat? Theoretically, this can be done by taking fat-precursor cells from the body and exposing them to chemicals that encourage them to develop into brown-fat cells. This has been done in mice and, sure enough, when injected back into animals the treated cells develop into

Hot masquerades as cold

If you shiver at the thought of exposing yourself to cold temperatures to lose weight, there may be a warmer alternative. Capsaicin, the substance that gives chillies most of their burn, seems to stimulate brown fat in a similar way to cold exposure. Mice given capsaicin as part of a high-fat diet increase their metabolic activity and do not put on weight. Men given a daily dose of capsaicin also boost their brown-fat activity and burn more calories. On the plus side, capsaicin is cheap, relatively safe and can be grown to order. On the down side, we still need large trials to prove that it can reliably stimulate brown fat and induce weight loss.

brown-fat cells. The same process may work in humans too.

There might be yet another option, however. A few years ago scientists discovered a third type of fat, called beige fat. This has a different origin from brown fat and, rather than growing in pockets, it is distributed throughout white fat. Most importantly, beige-fat cells contain the heat-generating protein UCP1.

Beige fat has several potential advantages over the brown variety. Exposure to cold not only activates beige-fat cells to generate heat but has the added bonus of increasing their number. That's one better than brown fat.

What's more, research in rodents shows that it is possible to turn white fat into beige. When exposed to cold, mice naturally produce chemical signals that act on immune cells called macrophages. In turn, these trigger white-fat cells to become beige. When mice are injected with these same chemical signals, it is as though they have been exposed to cold – white fat become beige fat and, best of all, the mice burn up more energy.

Master switch

We have also discovered a genetic 'master switch' that directs white-fat cells to stop storing energy and burn it instead – effectively turning them into beige cells. Researchers are using gene-editing techniques to explore the possibility of creating a gene therapy for obesity.

White, beige or brown, our understanding of fat cells and how to manipulate them is growing at an impressive rate. And while much still needs to be done if beige and brown fat are going to enter the weight-loss business, there's plenty to be optimistic about.

4

Get Inside Your Head

The nature of thought

Try, if you can, to imagine a life without thought. For a human being it wouldn't be much of an existence. Thoughts fill our every waking moment, and thinking comes naturally to us. We might say that thought is to human beings what flight is to eagles and swimming is to dolphins.

But it is one thing to think and quite another to understand what thought is. Just as eagles fly without any grasp of aerodynamics and dolphins swim without understanding fluid mechanics, so most of us think without having any insight into its nature.

So what is thought? That is a surprisingly difficult question to answer. Thought is an extremely varied and complex phenomenon. We can think about an incredible variety of things: objects, people, places, relationships, abstract concepts, the past, the future, real things and imaginary ones. We use thought to solve problems and invent things. We can think about nothing at all, and even think about thought itself.

To make some progress, we first need to lay down some definitions, because the term 'thought' can refer to three quite different things.

Choose your concept

In one sense, thought refers to a type of mental event. To think of something is to bring it to mind. In another sense it refers to a certain kind of mental faculty. Just as there are faculties associated with perception and language, so too there is a mental faculty – or perhaps faculties – associated with the capacity to think. In a third sense it refers to a certain kind of mental activity. Just as you can be engaged in the activity of looking or listening, so too you can be engaged in the activity of *thinking*.

Consider thought as a mental event. What are thoughts, and what distinguishes them from other kinds of mental events?

Suppose you are having a bonfire. You can see the flames and feel the heat. These are purely perceptual. But you may also find yourself wondering what would happen if the wind changed direction, or how combustion works. These events are prompted by perceptual experience, but are not themselves forms of perception. They are thoughts.

Think of an apple

The contrast between perception and thought can also illuminate thought as a mental faculty. In order to perceive an apple, say, there must be a causal connection between you and it. Light must be reflected from the apple and be processed by your visual system. No such connection is required

Don't think of a white bear

Attempting to control the direction of a stream of thought can be counterproductive. In a famous study, psychologist Daniel Wegner instructed one group of volunteers to think about white bears for a 5-minute period, and another group to not think about white bears for the same amount of time. The second group thought about white bears more than the first. This is known as the 'white bear problem': attempts to suppress a thought usually backfire.

Tim Bayne is professor of philosophy at Monash University in Melbourne, Australia, and author of *Thought: A Very Short Introduction*

to think about an apple. Another contrast between perception and thought concerns the range of properties that these faculties can acquaint us with. Perception provides us with access to only a limited range of properties. You can perceive that an apple is red, but you cannot perceive that the apple originated in western Asia, or that it has more genes than a human. You can, however, grasp these features of the apple in thought, for not only can you think 'That apple is red', you can also think 'Apples originated in western Asia' and 'The apple has more genes than a human'.

Mental activity

What about thought as a mental activity? Although thoughts can occur in isolation, it is perhaps more common for them to come in trains. Sometimes thoughts are related associatively, with one naturally and effortlessly leading to another as in a game of word association. Thoughts of Switzerland might trigger thoughts of skiing which might lead to thoughts of snow which might lead to thoughts of Christmas, and so on.

Although there is a certain delight to be had in following this kind of train of thought, the power of thinking arguably resides in something more systematic: the fact that it enables us to use evidence and logic.

Consider the chain of thought 'Socrates is a human', 'All humans are mortal' and 'Socrates is mortal'. The components of this train of thought are inferentially connected, for if the first two thoughts are true then so too is the third. Much of the value of thinking comes from our ability to organise thoughts into coherent trains to 'see' what follows from what. In other words, much of our interest in thinking concerns reasoning.

Having distinguished these various aspects of thought, we can now turn our attention to the nature of thought. What exactly is it?

It used to be believed that thought required some kind of non-physical medium – a soul or an immaterial mind. Modern theorists typically reject this view in favour of a materialist account, according to which thought involves only physical processes.

There are many reasons to believe this. Perhaps the most convincing is that fMRI scanners can detect thoughts as patterns of physical activity in the brain.

In one study, volunteers were asked to choose between two options – 'add' or 'subtract' – before being presented with two numbers on which to perform their chosen operation. From fMRI scans, researchers were able to tell with 70 per cent accuracy whether the subjects had decided to add or subtract, thereby reading their hidden intentions. Brain imaging is a long way from completely decoding the 'language of thought', let alone designing a machine that can read people's thoughts, but it does suggest that thoughts are material states.

Is language essential?

One contested question about the nature of thought concerns the role that language plays in it. There is a wide range of opinion on this topic. Some theorists hold that we think in language, whereas others hold that language plays no role in thoughts other than to allow us to communicate them. The truth is likely to lie somewhere in the middle.

One way into this debate is to consider what kinds of thoughts non-human animals can entertain. Many species have some capacity to track mathematical properties. Rats can be taught to press a lever a certain number of times for a food reward, and chimps can compare quantities quite accurately. Faced with a choice of two trays, each with two piles of chocolate chips on them, they can usually determine which tray has the most chips overall. Chimps can also grasp simple fractions. When shown half a glass of milk, they are able to point to half an apple and ignore three quarters of an apple in order to gain a treat. These skills presumably require something like thought.

Another area where thought-like representations have been found in non-human animals is in the understanding of psychological states. Primates, at least, seem to be able to determine what others can see – and thus, perhaps, what they know – on the basis of what they are looking at. They will follow the gaze of others to locate the object of their attention and will remove food items from the line of sight of other animals.

Augmented powers

It seems clear that a number of non-human species possess the capacity for at least primitive forms of thought. However, it is unlikely that any other species possesses the capacity to entertain the range of thoughts of which we are capable. What accounts for the sophistication of human thought? The answer appears to be related to language.

We can see how language might augment the powers of thought by considering an experiment involving Sheba, a chimpanzee trained to use numerals to represent items. Sheba was offered two plates of food, one large and one small. To obtain the larger plate, she had to point to the smaller one. Although she understood the rule, she wasn't able to override her instinct to point towards the larger plate – until the plates were covered and numerals representing the number of treats were placed on top of them.

The use of symbols allowed Sheba to transcend her normal abilities and disengage her thought from perception. This 'decoupling' is a striking feature of human thought, and may be facilitated by – and perhaps even require – the use of symbols. Language also facilitates thought in other important ways. By putting thoughts into language we are able to take a step back and subject them to critical evaluation.

Active or passive?

Another key question that arises from considering thought as an activity concerns the kind of control we have over it. Is thinking an intentional and controlled activity, or is it largely passive?

Suppose that I ask you why democracies tend not to wage war against other democracies. (It is often said that democracies have *never* waged war on one another, but that is not true.) If you have not already considered this question, you may need to think about it.

What precisely does that involve? If your experience is anything like mine, you simply put the question to yourself and wait for something to spring to mind. There is no rule that you can consciously follow in order to generate the required thoughts.

Where thoughts come from

On the whole, thinking often doesn't seem to extend much beyond putting questions to yourself and waiting for your unconscious to answer. The role of consciousness in such cases seems to be that of a minder whose job is to ensure that one's train of thought doesn't wander off-topic. So although we have some conscious control over the direction of our thoughts, it is far from unlimited.

On the shoulders of giants

One very distinctive feature of human thought is that it occurs in a social environment. We are born into a community of thinkers, and we learn to think by being guided by those who are experts. Indeed, childhood is an extended apprenticeship in thinking. Cultural transmission allows the best thoughts of one generation to be passed on to the ones that follow. Unlike other species, whose cognitive breakthroughs usually have to be rediscovered anew by each generation, we are able to build on the thoughts of our ancestors. We inherit not just the contents of their thoughts, but also methods for generating, evaluating and communicating thoughts.

The potential of human thought is clearly very great. It is not limited in the way our physical and perceptual abilities are. We cannot see or visit distant tracts of space and time, but we can think about them. But are there limits to what our minds can grasp? The idea that certain aspects of reality are beyond us might seem implausible. There doesn't seem to be any aspect of the world that we cannot think about.

Limits to thought

Is there any reason to take the possibility of cognitive limits seriously? There is. Consider the cognitive limitations of other species. Chimpanzees might be able to think about a range of things, but it is doubtful whether they can think about quantum mechanics. Perhaps that is one of the limitations of lacking language. But if there are aspects of reality that are inaccessible to the minds of other thinking species, perhaps there are also aspects of reality that are inaccessible to ours.

It is one thing to grant that some aspects of reality lie beyond our grasp, but quite another to identify what they might be. Is it possible to demarcate the borders of human thought?

You might worry that the question is absurd on the grounds that if a certain thought were unthinkable (by us) then by hypothesis we couldn't think about it, let alone know that it was unthinkable. But there is nothing paradoxical about attempting to determine where the limits of thought lie. The key involves distinguishing thinking about a thought from actually thinking it. Just as we can know what we don't know – the known unknowns – so too we might be able to think about some of what we cannot think: the thinkable unthinkable.

Wherever the boundaries of human thought lie, there is no doubt that we are very far from having reached them. There are thoughts – deep, important and profound thoughts – that no human being has yet entertained. Thought has taken us a long way; who knows where it will lead?

What's going through your mind right now?

Your brain is peerlessly creative, but also home to wildly veering trains of thought, idiosyncratic memories, strange obsessions and delusional beliefs. Only you know what goes on in there. You may consider its workings to be perfectly ordinary or wonderfully weird, but how would you know? After all, none of us has ever experienced being inside the brain of another. What counts as a 'normal' mind? Is there even such a thing?

What, for example, occupies the thoughts of most people most of the time? Rumour has it that men think about sex every 7 seconds. That's almost certainly untrue. Male students given clickers and told to click whenever they thought about sex responded an average of 19 times a day. For women, the average was around 10. Thoughts of food were just as common, numbering 18 a day for men and 14 for women. And sleep registered a respectable 10 for male students and 9 for females.

Other studies show that pleasurable subjects dominate most people's spontaneous musings – not just sex, food and sleep but also things like socialising and shopping. It seems humanity is united in its humdrum cogitations. Lofty thoughts are uncommon even among intellectuals. But dark thoughts are relatively rare too. Unless directly confronted by death, for example, most of us almost never think of it.

Cognitive doodling

A morbid obsession with death affects around 15 per cent of people, but obsessive thinking in general is quite common. We tend to characterise wandering thoughts as random, or loose chains of association, but if you find your mind constantly meanders back to familiar territory you are not alone. Like cognitive doodling, obsessive or ritualistic thinking might just be a way of occupying the idling mind. However, such thoughts once carried an evolutionary advantage, as they prepared us for dealing with future risk. That would explain why they are often to do with possible threats, such as uncleanliness.

Of course, much of the time your thoughts are directed by what you are doing, whether it be working, socialising or watching TV. At least, they would be if you could just pay attention. Most people think they are more prone to mind-wandering than average. In fact, what's normal is hard to judge. In experiments to find out, people are asked to read extracts from books such as Leo Tolstoy's *War and Peace*. Interrupting them to ask their thoughts at random intervals reveals we spend anywhere between 15 and 50 per cent of the time with our head in the clouds.

Attention span

Don't worry either if you have a short attention span. There's a lot of variation in people's ability to stay focused on one task. It is poor in kids, perhaps because the developing brain has yet to master control over areas that process incoming sensory information. It then improves until the age of 20, when it plateaus until middle age, before diminishing once again. The average attention span could be shockingly short. A report by Microsoft put it at just 8 seconds. However, there's no evidence that technology is making us worse at concentrating.

What about memory? Obviously, the suite of memories you have is unique to you. But you may wonder why your brain tends to favour certain kinds of memories. Long-term memories divide into two main types. Semantic memories record facts, such as a train timetable. Episodic ones are about events we have experienced, such as a particular train journey. And true to the stereotype, women tend to have better episodic memories than men. What's more, with semantic memory, men tend to remember spatial information better, whereas women generally perform better at verbal tasks, such as recalling word lists. Personality type seems to be a factor, too: people open to new experiences tend to have better autobiographical memory.

Talk to yourself

The quirks of the average human mind are stranger than you might think. Even hearing voices isn't that odd. Some 60 per cent of us experience 'inner speech', with our everyday thoughts taking a back-and-forth conversational quality. What's more, between 5 and 15 per cent of us hear outside voices, even if only fleetingly or occasionally.

About 1 per cent of people with no diagnosis of mental illness hear recurring voices. Around the same proportion of the population is diagnosed with schizophrenia, challenging the assumption that the two are related. In fact, scans reveal little difference between the brains of people who haven't been diagnosed with mental illness, but do hear voices, and those who don't hear voices.

Ageing affects the recall of personal experiences more than that of facts. It's not that our brains are overloaded – our memory capacity is practically unlimited. Rather, gradual changes in brain structure are to blame. Whatever your age though, memory is mostly about forgetting. So you shouldn't be overly concerned if your memory seems to move in mysterious ways. It is a personal thing, reflecting what you consider to be important.

Our memories tell us stories, too. This 'confabulation' is a symptom of some memory disorders, whereby people have false recollections. But the rest of us do it too. Experiments show, for instance, that when people are forced to make a random decision they later invent a narrative to explain it.

One explanation is that this helps us make sense of a world that bombards us with information, and gives conscious rationale to decisions we make for unconscious reasons. Our lies may also be more self-serving: by lying to ourselves, we lie better to others too.

Delusion-like beliefs

Even your strange beliefs don't set you apart. Once upon a time, believing impossible things would have been seen as a sign of mental imbalance. Today we know that most people hold at least one 'delusion-like' belief. These are mild versions of beliefs that could get you diagnosed with a mental illness – that your family have been abducted and replaced by impostors, for example. Most of us seem untroubled by them.

So if you are delusional, can't concentrate for more than 8 seconds, forget people's names as soon as you are introduced, and can't stop thinking about sex or food, congratulations. You have a normal human mind.

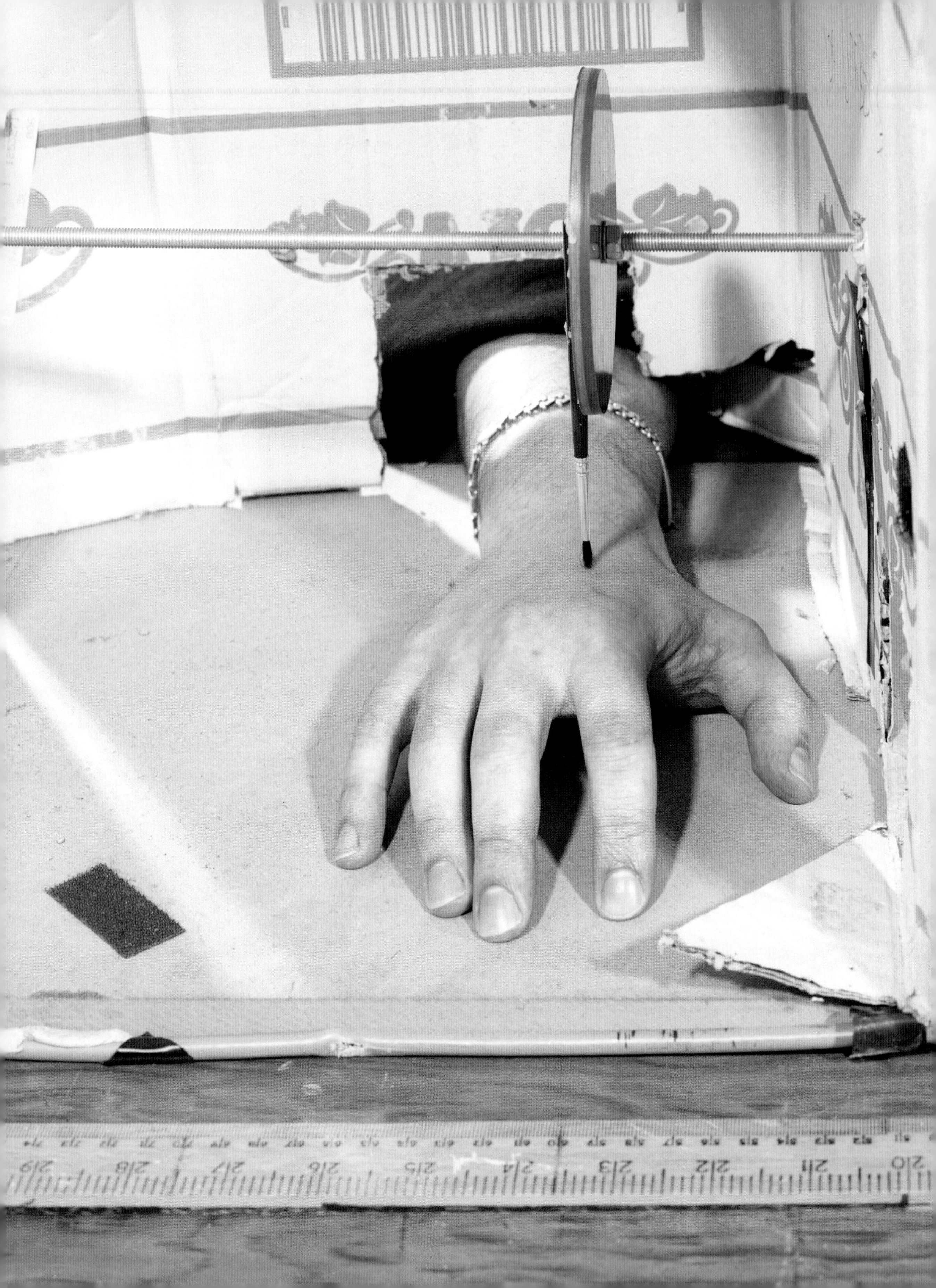

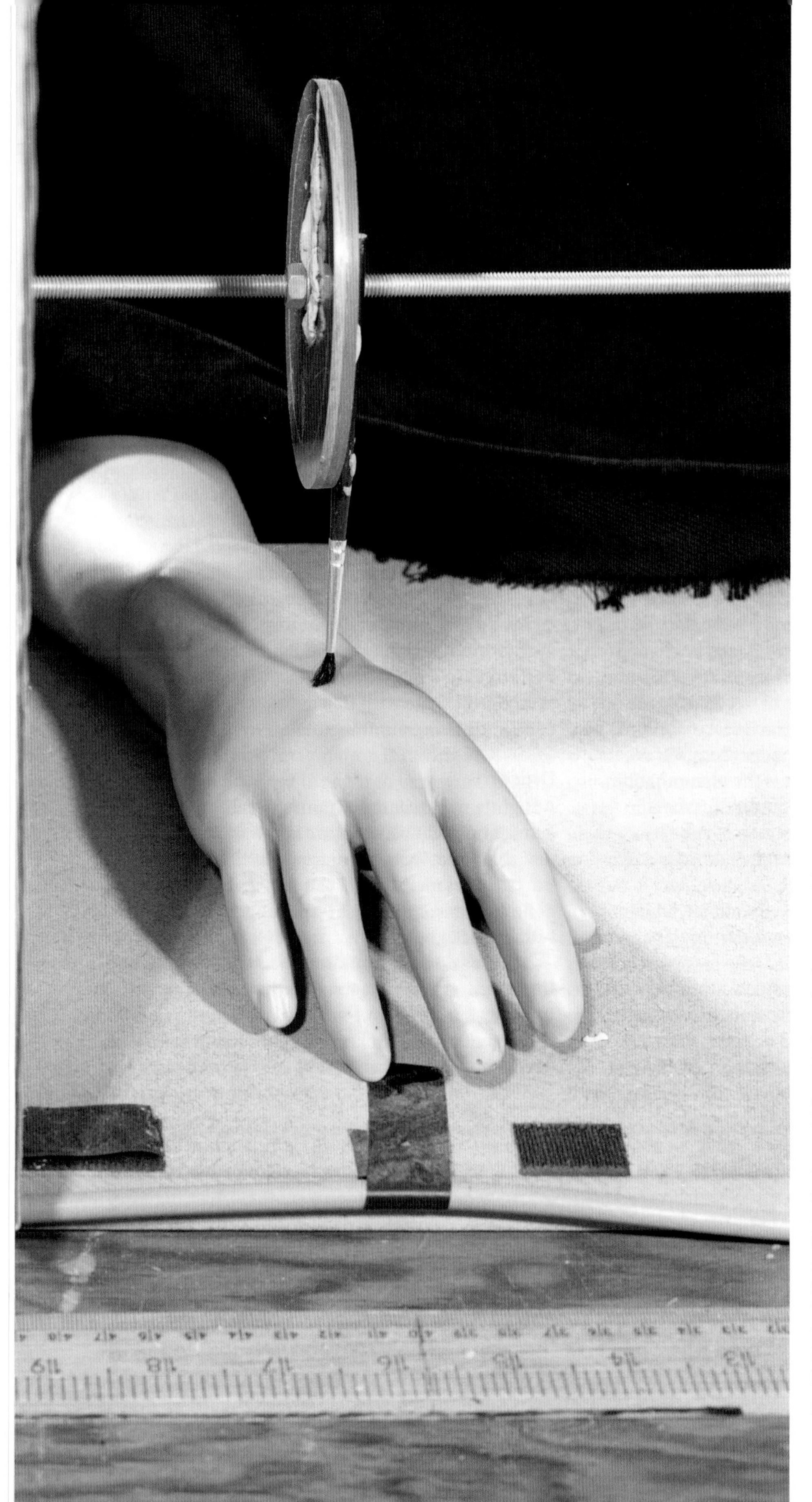

Where do I begin and end?

Your sense of self is such a concrete thing you may be shocked to hear that it is probably an illusion created by your mind. You cannot even be sure where your bodily self begins and ends – as the rubber-hand illusion reveals. Your real hand is placed behind a screen: all you see is a dummy hand. Then your real hand and the fake one are stroked at the same time with the same motions. After a while your brain incorporates the dummy hand into its body schema and you start to feel that it is yours. If somebody makes to stab it, you will flinch and your brain will prepare to feel pain. Such is the power of your senses that the visual and tactile signals to your brain override your knowledge that the fake hand is not yours.

Credit: Daniel Stier

Why your memory is not only for remembering

Welcome, time traveller. You may be physically stuck in the here and now, but mentally you know no bounds. All you have to do is recall an event in your past and you have performed a feat that, as far as we know, is uniquely human.

We are all collections of memories. They dictate how we think, act, make decisions, and even define our identity. Yet memory is a puzzle. Why do we remember some things but not others? Why does memory play tricks on us? And is memory one thing, or many?

On that latter point, at least, we can be clear. Memory is built of many different components. Broadly speaking there are three types: sensory, short-term and long-term.

Memory relay

Sensory memory is the most fleeting; it briefly retains incoming information before shunting relevant pieces into short-term memory. This is also a temporary holding pen, capable of storing roughly seven items of information for up to 20 seconds. It is where, for example, you hold the digits of a PIN number, or manipulate information as you solve a puzzle. From short-term memory, salient information gets transferred to the brain's long-term storage facility, where it can remain for the rest of your life.

Long-term memories also come in different flavours. There's semantic memory, which is where you store facts; procedural memory for skills such as riding a bike; and episodic or autobiographical memory, which records events from your life.

It is this type of memory that is most intimately tied to our sense of being human. Events such as graduation day, a first kiss, or the day we lost somebody special are autobiographical memories, and an essential part of who we are.

When thinking about this sort of memory, it is natural to envisage it as a mental diary – the private book of you. To relive your first day at school, say, you simply dust off the cover and turn to the relevant page. But there is a problem with this common-sense view. Why are the contents so unreliable? Why do we forget key details, mix things up and even 'remember' events that never actually took place?

Such flaws are puzzling if the purpose of memory is simply to record your past. But they begin to make sense if it is for something else as well. That is what memory researchers are now starting to realise: memory is what allows us to imagine the future.

The first inkling that this may be the case came out of studies of people with a type of amnesia that destroys their autobiographical memories. These people often struggle to make plans, as if being robbed of their past has also robbed them of their future.

Brain scans support the idea, suggesting that every time we think about the future, we tear up pages of our autobiographies and stitch together the fragments into a new scenario. Memory allows us to project both forward and back, an all-purpose 'mental time travel machine'.

I don't remember that!

Our autobiographical memories are incomplete in other ways. Some periods of our lives generate heaps of memories, while others receive relatively patchy coverage. Why do we remember some events but not others?

Young children are notorious amnesiacs. Our brains do start remembering at a young age, learning simple associations even before we are born, but we cannot consciously remember specific events from before the age of about three. Even then, we are hard-pressed to remember much from before our sixth birthday.

Three different factors have emerged to explain this hazy recall. One possibility is that the neural pathways are not mature enough between the hippocampus – where memories are consolidated – and the rest of the brain, so experiences are never cemented into long-term storage. Language development also matters, because words provide a kind of scaffold on which we hang our memories. Children don't remember an event until they have learned the words to describe it.

Who am I?

A sense of our own identity is also crucial. Children who can recognise themselves in a mirror – a sign that they have developed a sense of self – are capable of recalling certain events a week later, while toddlers who fail the mirror test draw a blank (see 'What's going through your mind right now?', p 88).

As we grow up, autobiographical memory becomes more active and eventually goes into overdrive, peaking in early adulthood. We are more likely to remember things that happen in this period than from any other time in our lives. This 'reminiscence bump' may be the result of anatomical changes to the developing brain, or maybe the fact that our brains feel emotions much more keenly during adolescence and early adulthood: memories linked to intense feelings stick in the mind for longer. Or perhaps it is simply down to the fact that many important landmarks and rites of passage in our lives fall within this period.

We even anticipate the bump. When young children are asked to imagine their future life stories, most of the events take place in young adulthood. The finding dovetails with the idea that memory and foresight share the same machinery in the brain.

As we get older, memory fades and future gazing becomes less important. But many older people still retain the ability to unshackle themselves from the here and now and time travel back to life-defining events. Autobiographical memory may not be reliable, but it remains an indispensable part of who we are.

False memories

It is very easy to lead memory astray. If somebody who witnesses a traffic accident is later asked whether the car stopped before or after the tree, they are likely to remember there being a tree at the scene even if there wasn't. This 'misinformation effect' can even be used to persuade people that they took part in fictitious events that they really ought to know didn't happen, such as a balloon ride or visit to Disneyland. For some events, however, memory is very robust. The death of Princess Diana and the events of 9/11, for instance, create strong and vivid memories that persist for decades. This is what has been termed 'flashbulb memory'.

So you think you're in charge of you?

Think you know what's going on in your mind? Think again. Much of our mental life happens out of sight, in a place that was once considered the cesspit of our basest desires.

That place is the subconscious (or unconscious), a dark and disreputable corner of the mind first conjured up by Sigmund Freud as part of his now-discredited theory of psychoanalysis. Freud and his followers saw the unconscious as little more than an emotional and impulsive force in constant conflict with the more logical and detached conscious mind.

This is a view that modern neuroscientists definitely don't share, but they do agree with Freud on one thing – that our brains have an uncanny knack for working stuff out, with no need for conscious involvement. Subconscious thought processes seem to play a crucial role in many of the mental facilities we prize as uniquely human, including creativity, problem-solving, memory, learning and language. For some tasks, it is superior to rational, conscious deliberations.

Some scientists go so far as to believe that it is responsible for the vast majority of our day-to-day activity and that we are nothing more than zombies guided by our unconscious. That's an extreme position, but there's no denying the power of the unconscious mind.

Modern study of the subconscious began with a famous experiment in the 1980s. Benjamin Libet at the University of California, San Francisco, asked people to press a button whenever they wanted to. The exact time they decided to act was recorded on an ultra-precise clock, and the volunteers also had electrodes placed on their scalp to measure electrical activity in their brain.

This set-up revealed that neuronal activity preceded people's conscious decision to press the button by nearly half a second. More recently, a similar experiment with an fMRI scanner found stirrings in the brain's prefrontal cortex up to 10 seconds before someone became aware of having made a decision to act.

These results are sometimes interpreted as disproving the existence of free will – though they could equally mean that we do have free will, but it is exercised by our unconscious mind (see 'Do you have free will?', p 46). Whatever it said about free will, the experiment showed that there are important mental processes bubbling under the surface of awareness.

Breaking through

Accessing those processes requires some clever experimental techniques. One is called 'masking'. Volunteers are shown a word for a few tens of milliseconds, and then shown an image that masks the word and stops the subject from consciously perceiving it. As the time before the mask appears is increased, the word suddenly pops into consciousness – accompanied by characteristic activity visible on a brain scan.

The masked word usually arrives into consciousness when the interval is around 50 milliseconds – less if the word has emotional significance, which makes it more attention-grabbing. Neuroimaging confirms this, revealing widespread brain activity once the word is consciously perceived.

Masking experiments have shown that unconscious information can influence conscious thoughts and decisions. For instance, people shown the masked word 'salt' are then more likely to select a related word from a list, like 'pepper'.

Experiments like these have changed our views about the relationship between conscious and subconscious, putting the latter firmly in charge.

Our unconscious brain constantly monitors the world and, when the input becomes important enough, shoves it into conscious awareness. Think of consciousness as a spotlight, with the subconscious controlling when to turn it on and where to direct the beam.

The unconscious is also the brain's autopilot. Much of what we do in day-to-day life happens beneath conscious awareness. That includes routine tasks such as swallowing and breathing, but also skills like driving, playing golf or touch-typing. When we start to learn these skills they take up all of our attention, but with practice they are delegated to the subconscious, freeing up attention to be focused elsewhere.

The unconscious is also the place where snap judgements are made. If you have ever fallen in love at first sight or felt an irrational distrust of a stranger on a bus, that's your unconscious mind doing its thing. Surprisingly, these snap judgements are often pretty accurate. Experiments show that we can judge somebody's confidence, sexuality, economic success and political affiliation after seeing them for just 2 seconds.

No sleep for the unconscious

Our unconscious minds are even purring away while we sleep. If you have ever had the uncanny experience of waking up just before your alarm goes off, that's your unconscious mind marking time while your conscious one slumbers.

But there's more to the unconscious than menial tasks. Non-conscious thinking may actually work better in some situations where you might imagine rational, conscious thought to be best. People who have to make a difficult choice based on large amounts of hard-to-compare information – for example, which apartment to rent or mobile phone contract to take out – often do better if they don't actively think about it. In these situations, it seems the unconscious does a better job of weighing up the pros and cons.

In common with the default network (see 'Why does your mind always wander?', p 94), the unconscious mind's way of processing information may also be important for creativity.It brings together disparate information from all over the brain without interference from the brain's goal-directed frontal lobes. This allows it to generate novel ideas that burst through to consciousness in a moment of insight. Such non-deliberate deliberation may also explain those 'aha!' moments when the answer to a problem seems to come from nowhere, as well as times when an elusive word or name comes to mind only after we stop groping for it.

Seeing without sight

One spooky way to access the unconscious mind is to study people with a condition called blindsight. Brain damage following an injury or stroke very occasionally leaves people with no conscious awareness of being able to see, but does not affect their ability to navigate a cluttered room, identify objects and respond to emotional facial expressions. This suggests that while they may not consciously see a stimulus, they are able to subconsciously process what they have seen and respond appropriately.

Why are we so prone to bad habits?

We humans are creatures of habit. Around 40 per cent of the things we do every day require no conscious thought whatsoever. That includes healthy behaviours such as brushing our teeth, but also unhealthy ones like smoking. Like so many things in life, habits are a double-edged sword.

Just why habits are so hard to make and break is a long-standing mystery. Even so, the prospect of mastering our habits has enormous appeal. Accepted wisdom suggests, for instance, that it takes 21 days to form a new habit or get rid of an old one. Unfortunately, there's little by way of evidence to back up such notions. But that is starting to change. Neuroscientists are building an accurate picture of what happens to brain circuitry when a new habit is formed, and hence – maybe – how to break the habits of a lifetime.

The first challenge in understanding habits is getting to grips with what one actually is. In the vernacular, habits are usually seen as undesirable – bad table manners or smoking, for example.

Scientifically, though, habits are defined broadly as any action performed routinely in a certain situation, often unconsciously. Once a habit is formed, it is like a program that runs on autopilot.

This process plays a vital part in making everyday life easier: if you had to give your full attention to brushing your teeth or the commute to work every time you did it, life would become overwhelming. In fact, as much as 40 per cent of our daily behaviour is habitual. When you engage in well-practised behaviours you are often thinking of something else.

'Waking up' halfway home

Habitual behaviours can be surprisingly complex. Have you ever got into your car intending to drive to the supermarket and 'woken up' halfway to work? Or perhaps driven home to a house you no longer live in? These slips of action are a manifestation of habitual behaviour.

Pathological habits?

Biting your nails might be impolite and unsightly, but it's not life-altering or life-threatening. Where do bad habits become more of a problem, such as addiction or obsessive-compulsive disorder (OCD)? There's evidence that people with OCD, Tourette's syndrome or drug problems have disruptions in brain circuits involved in habit formation. Research has found people with OCD to be more vulnerable to 'slips of action' – unintentionally executing habitual behaviour. Anorexia might also be an extreme form of habit.

With drug use, however, it gets more complicated, because the neurotoxicity of drugs also affects the brain. So while having a strong propensity for forming habits might make you more likely to become addicted, the drug itself might make you more prone to falling into habit traps.

All this makes sense from a practical perspective, but it also suggests that something changes in the brain when a conscious action turns into a habit.

After a rat learns to navigate a maze and begins to follow the same route out of habit, brainwaves slow down in an area of the brain called the striatum. This may indicate the creation of the habit. Activity in the striatum changes in a similar way when monkeys learn a new habit.

Importantly, the studies showed that cells within the striatum fire in this way at the beginning and end of a behaviour, as if signalling when the autopilot program is turned on and off. You might think of this as the brain's equivalent of 'chunking', the method for remembering lists of things by dividing them into discrete chunks. If you've ever been interrupted while reciting a phone number, you probably had to start over, because you only knew the full number as a sequence of chunks.

The chunking of behaviour is what allows us to avoid wasting valuable brain power on routine activities. But it obviously has a downside: your brain can also habituate unhealthy or unwanted behaviours.

Most habits start off as positive, goal-directed behaviours: you want a tidier bedroom, so you make your bed each morning. Repeated often enough, it becomes automatic. A bad habit like nail biting can also start out as goal-directed, for example to relieve stress. But it, too, can become habitual, and soon enough you are biting your nails without being aware that you're doing it.

Annoyingly, our brains do not distinguish between good and bad habits. Evidence for this comes from studies of willpower. Willpower comes in limited supply; the more we use it during the day the more it gets depleted, which means we're more likely to give up on later attempts.

When willpower is depleted – in times of stress or exhaustion, for example – we fall back on habits, whether good or bad. It's perhaps not surprising that during exams, students find that habits like unhealthy snacking increase. But paradoxically, good habits, like reading or exercising, also increase.

We also know that habits are triggered by certain cues or environments. If you give people popcorn while watching videos in either a cinema or a conference room, for example, they generally eat more popcorn in the cinema, even if the popcorn is stale. It could be environmental cues that kick off brain signals that tell the striatum to initiate that habitual behaviour.

Kicking bad habits

Understanding what habits are, how they form and what triggers them could help to answer the million-dollar question: how to get rid of bad habits and retain or create good ones.

The link between habits and our environment is why the best times to break habits or deliberately create new ones are when we go away, change jobs or move house. Another tip is not to worry about little slip-ups. One study followed about 100 people as they tried to form new habits. It found no long-term consequences to slipping up for a day here or there. And it gets easier. Willpower is like a muscle; although it can get depleted, it also gets stronger with practice.

How long do you need to practise for? Folk wisdom suggests that it takes three weeks for habits to form, but scientific studies produce a figure of more like 10 weeks, ranging from 18 to 254 days. But once you have made a good habit, it can persist for a lifetime.

Don't believe everything you think

We humans have two ways of thinking. The first is swift and intuitive. Called system 1, it effortlessly solves the myriad problems we face every day. We engage system 2 only when we meet complex mental tasks. Sadly, for us, this cognitive machinery makes predictable errors

Question 1

A table tennis bat and ball cost £1.10 in total. The bat costs £1.00 more than the ball. How much does the ball cost?

Answer

Many people intuitively answer 10 pence for the ball. But that would mean the bat costs £1.10 and the total cost £1.20. The correct answer is 5 pence for the ball and £1.05 for the bat.

Bias

The intuitive answer is 10 pence – it just pops into your mind. To get past this system 1 answer, you need to invest a little effort by using system 2 to do the maths. Some people do this automatically, many do not. Notice that even when you know the solution, 10 pence still looks an attractive answer

Question 2

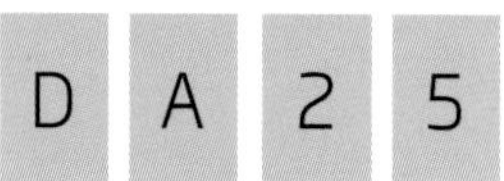

Four cards have a letter on one side and a number on the other. Which two cards should you turn over to show that the following statement is true: 'If there's a D on one side, there's a 5 on the other'?

Answer

Most people choose D and 5. But the statement says nothing about which letters might be on the reverse of a 5, so turning that over does not help. The correct answer is D and 2 because if there's a D behind the 2 then the statement is false

Bias

This reveals confirmation bias. We think we are weighing up alternatives rationally but we've already chosen a favoured option (D and 5 are best) and we want to confirm it. In fact, the best way to test such an idea is to try to disprove it

Question 3

Aunt Mary phones you distressed after watching a series of news reports about burglaries on the other side of her town. Her home is in a safe suburb with an active neighbourhood watch and has a burglar alarm. Do you calm her down or advise her to beef up security?

Answer

The best option here is to calm her down

Bias

System 1 follows simple rules, or heuristics. The 'availability heuristic' pushes us to assess the relative importance of things not on the evidence but on what comes most easily to mind. And what we usually remember is what's been in the media. Aunt Mary's home is no more threatened than it was a month ago, but those news reports made her think it is

Question 4

You have a candle, numerous thumbtacks, and a box of matches. Using only these items, mount the candle to a wall.

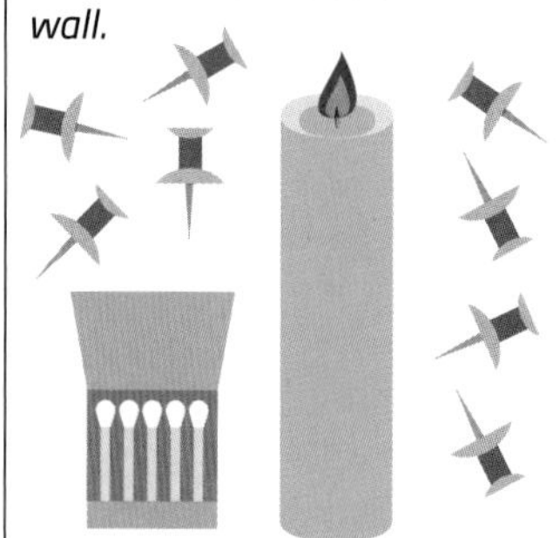

Answer

Empty the matchbox tray and pin it to the wall. Stand the candle on the tray. Strike a match and light the candle. Job done

Bias

If you think a matchbox tray is only good for holding matches, then you are displaying 'functional fixedness'. Usually it's a good shortcut for, say, finding the right tool for a job, but it can also limit your problem-solving abilities

Question 5

Linda is 25 years old, very bright, single and outspoken. When studying philosophy at college she was deeply involved in social justice issues and attended environmental protests. Rank the following statements from most to least likely:

A Linda works in a bookstore and takes yoga lessons
B Linda is a bank teller
C Linda is a bank teller, and is active in the environmental movement

Answer

There is no right answer, but there is a wrong one: b must come before c

Bias

Statistically it is much more likely that Linda is a bank teller than a bank teller plus an active environmentalist. Placing c above b is an example of the 'conjunction fallacy'. It places system 1 intuition above the logic of probability

Question 6

A father and son are in a car crash. The dad is killed and the son rushed to hospital. As he's about to go under the knife, the surgeon says: 'I can't operate – that boy is my son!' Explain.

Answer

The surgeon could be the boy's gay second father. But it's far more likely that the surgeon is his mother

Bias

Most people asked this question don't get the answer: it's an example of 'stereotypical bias'. For speed of access, our memories categorise people, places, things, and associate beliefs and expectations with them. The stereotypes formed in this way are often simplistic and wrong

Question 7

You're shopping for jeans and find the perfect pair. But you're stunned by the price tag: £150. Too much! When you get home you go online and find the same pair for £100. You buy them and bask in the glow of a good deal. But was it a bargain?

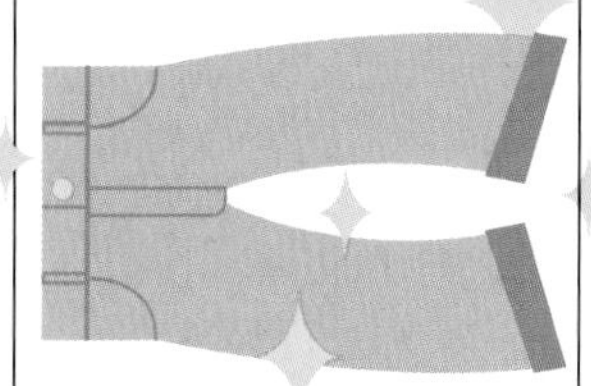

Answer

Jeans cost a fraction of £100 to make. So the answer is no

Bias

Humans are prone to the 'anchoring effect': we tend to rely too much on one piece of information, which colours subsequent decisions. The £150 anchor made the £100 price seem like a bargain. Many shops exploit anchoring by offering discounts on overinflated prices

Question 8

There is a patch of lily pads on a lake. Every day, the patch doubles in size. If it takes 48 days for the patch to cover the entire lake, how long would it take for the patch to cover half the lake?

Answer

47 days

Bias

This is similar to question 1. The intuitive answer is 24 because 'half the pond will be covered in half the time'. Now engage system 2 and think backwards. If the lake is covered in 48 days then a day earlier the lily patch would have covered half that area

How smart people can be stupid

'Earth has its boundaries, but human stupidity is limitless,' wrote Gustave Flaubert. The French novelist saw stupidity everywhere, from middle-class gossip to the lectures of academics. Colourful fulminations about his dunderheaded peers filled his letters to Louise Colet, the poet who inspired *Madame Bovary*. Not even Voltaire escaped his critical eye. Consumed by this obsession, he devoted his final years to collecting material for an encyclopaedia of stupidity. He died before it was complete, perhaps of exasperation.

Documenting the extent of human stupidity may itself seem a fool's errand, which could explain why studies of human intellect have tended to focus on the upper end of the spectrum. And yet, the sheer breadth of that spectrum raises many intriguing questions. If being smart is such an advantage, why aren't we all intelligent? Are there drawbacks to being clever that sometimes give the less sharp the upper hand? And why are even the smartest people prone to – well, stupidity?

Abstract reasoning

Attempts to study variations in human intellect tend to focus on tests that put a single number on it. IQ is widely acknowledged as an imperfect measuring stick. Nonetheless, it does capture something meaningful. It is perhaps best seen as a measure of abstract reasoning: the higher your IQ, the easier it is to grasp and master concepts such as calculus. People with an IQ of 120 find it quite easy; those at 70 have little chance of ever doing so. IQ also seems to predict academic and professional success.

Various factors determine where you lie on the IQ scale. About a third of the variation in intelligence is down to the environment in which we grow up – nutrition and education, for example. But the larger contribution, more than 40 per cent, comes from your genes.

Low IQ is clearly a handicap in some aspects of life, which raises the question of why low IQ genes persist. One possibility is that superior brain power comes at a cost. Unfortunately, evidence is in short supply. One of the only studies to report a downside to intelligence found that soldiers with higher IQs were more likely to die during the second world war. The effect was slight, however, and other factors might have skewed the data.

Alternatively, the variation in our intelligence may have arisen after civilisation eased the challenges driving the evolution of our brains. As human societies became more collaborative, slower thinkers were able to piggyback on the success of those with higher intellect. Some have claimed that a person plucked from 1000 BC and placed in modern society would be highly intelligent by today's standards.

This idea has some supporters, but the evidence is shaky. We can't easily estimate the intelligence of our distant ancestors, and average IQ has in fact risen slightly in recent decades – the famous 'Flynn effect'.

Perhaps the problem is putting too much emphasis on the supposed benefits of IQ. People with low IQ scores can still do clever things such as speak multiple languages and even engage in complex financial trades. Conversely, high IQ is no guarantee that a person will act rationally. IQ also doesn't immunise against the kind of irrational, illogical behaviour that most people would happily label 'stupidity'. You really can be highly intelligent, and at the same time very dumb. In fact, stupidity is most dangerous in people with high IQ since they often occupy positions of greater responsibility. What can explain this apparent paradox?

One idea comes from cognitive scientist Daniel Kahneman, who won the Nobel prize in economics for his work on human behaviour. Economists once assumed that people were inherently rational, but Kahneman discovered otherwise. When we process information, our brain can access two different systems: deliberation or intuition. IQ measures only the first of these, yet our default position in everyday life is to use intuition.

Ignorance ain't bliss

One major cause of stupidity is lack of knowledge rather than lack of intelligence. People who are ignorant about something often display preposterous overconfidence in their own abilities. The classic case was a novice crook called McArthur Wheeler, who pulled two bank jobs in Pittsburgh in full view of surveillance cameras. When police showed him the tapes he was incredulous. 'But I wore the juice!' he protested. Wheeler thought that rubbing his face with lemon juice – aka invisible ink – had made him invisible to cameras. Many similar examples of this 'Dunning-Kruger effect' have been documented. In a nutshell, it says that incompetent people's incompetence often causes them to fail to recognise their own incompetence.

Intuition is useful, providing cognitive shortcuts that cut through information overload and guide us to quick decisions. They include cognitive biases such as stereotyping, confirmation bias, and resistance to ambiguity – the temptation to accept the first solution to a problem even if it is obviously not the best. But if we rely on them uncritically, they can derail our judgement (see 'Why we're all biased', p 108).

Consider the following question: fish and chips cost £1.80. The fish is £1 more than the chips. How much are chips? The intuitive answer is 80p. But it's wrong (it's 40p).

Metacognition

An inability to recognise or resist these biases is often at the root of stupid decisions. Because it has nothing to do with IQ, susceptibility to these biases has inspired another measure of intelligence called the rationality quotient (RQ).

RQ also measures 'risk intelligence', which defines our ability to assess probability. For example, we tend to overestimate our chances of winning the lottery and underestimate the chance of getting divorced. Poor risk intelligence can cause us to choose badly without any notion that we're doing so.

Unlike IQ, RQ isn't down to your genes or nurture. More than anything, it depends on something called metacognition, which is the ability to assess the validity of your own knowledge. People with high RQ have acquired strategies that boost this self-awareness. One simple approach is to take your intuitive answer to a problem and consider its opposite before coming to the final decision. This helps you develop keen awareness of what you know and don't know.

So however clever you are, never forget that you're only one simple mistake away from being a clever idiot.

Why we're all biased

Conjure up an image in your mind of the two most recent ex-residents of the White House, George W. Bush and Barack Obama. Chances are you feel warmly towards one of them and dislike the other; the one you dislike you probably think was one of the worst presidents of all time.

Bush spilt the electorate right down the middle; around 50 per cent of Americans loved him and the rest hated him. Obama was even more divisive. At one point 81 per cent of Democrats thought he was doing a good job but only 13 per cent of Republicans agreed.

How can so many people make a judgement about the same president and come to such different conclusions? The obvious explanation is that they are biased – by their political tribe, by the media, by their friends and family and much else.

This obvious explanation is correct. But who, precisely, is biased? It depends who you ask. Those who think Obama was a great president think the conservatives, and their media, are the biased ones. Those who don't, think it's the liberals. In fact, they are both right.

A veil of prejudice

As any psychologist will tell you, pretty much everything you think and do is coloured by biases that you are typically totally unaware of. Rather than seeing the world as it is, you see it through a veil of prejudice and self-serving hypocrisies.

To get a handle on this, think about your own opinion of Bush, Obama, or indeed Donald Trump. You probably believe your view to be an honest and objective assessment based on a range of evidence. Perhaps you'll grudgingly acknowledge that you're inclined to cut some slack because you are liberal/conservative, but then reassure yourself that being liberal/conservative is the only rational choice, so that's OK.

You have just experienced the illusion of naive realism – the conviction that you, and perhaps you alone, perceive the world as it really is, and that anybody who sees it differently is biased. This conviction is very hard to escape.

If, at this point, you are thinking: 'Yeah, right, that might be true of other people, but not me,' then you have fallen foul of yet another aspect of the illusion: the bias blind spot. Most people will happily acknowledge that such biases exist, but only in other people.

Below the radar

Why are we so blinkered? The problem is that our biases – which form and solidify in childhood and early adulthood – operate below the radar, in our subconscious. It is not that people do not look inwards to question their own judgements and beliefs. Many do. But their biases are not consciously available for inspection, so they leap to the conclusion that their beliefs are correct and based on rational reasoning.

Many of the biases are a harmless variant of the positive illusions we routinely entertain in order to shelter our fragile egos from reality, such as the 'better-than-average' effect that convinces us that we're above average on a range of desirable abilities (see 'Why you're good at everything', p 52).

Others are more serious. Few people believe that they are racist, and their beliefs are honestly held, and yet time and again they are betrayed by their brains.

These 'implicit biases' can be revealed by laboratory tests which flash up quick-fire images of faces followed by words such as 'good' and 'bad'. The task is to decide whether the word is positive

Reminisce with caution

It's not just your opinions and knowledge that diverge from reality. Memories, too, are deeply suspect. Most of the evidence for this comes from 'false memory research', where psychologists deliberately plant fake memories into people's heads. In one famous experiment, doctored photographs and fake parental testimony were used to convince people they had been taken on a fictitious hot-air balloon ride as a child. Existing memories can be altered too: when asked about their memories of the death of Princess Diana, including whether they had seen 'the footage' of the actual crash, nearly half of people said they had, even though no footage exists.

or negative. As a rule, people are quicker to identify positive words after seeing a white face and negative words after a black face. These tests also expose unconsciously negative attitudes to gender and homosexuality.

Alternative facts

While opinions are obviously ripe for bias, facts are also at its mercy, with people adept at interpreting the world to fit with their existing beliefs. For example, environmentalists interpret the fact that the majority of scientists and governments are convinced that humans are changing the climate as open-and-shut evidence that we are. People on the other side just see a conspiracy. No amount of new information will change those positions, and yet on the whole, both camps sincerely believe their views are unbiased and rational.

Self-serving slant

Similarly, we seek out information that fits with our beliefs and ignore or dismiss information that doesn't. This 'confirmation bias' has been shown time and again, for example in experiments in which people are asked to read a range of evidence about a contentious topic such as capital punishment. Even when exposed to arguments on both sides, most people interpret the evidence in a self-serving way, accepting the data that supports their pre-existing views and dismissing or ignoring the rest. The scary thing is that they have no awareness of doing it. Similarly, confronting people with new information that contradicts their beliefs more often than not ends up hardening their position.

We also apply 'hindsight bias' to simplify and justify the past. Ask somebody why they bought the car they did and they will say something like 'superior fuel efficiency', when in fact they just liked it better for reasons they cannot explain. It obviously hardly matters why somebody really bought a car, but our ability to live in different factual universes over such issues as climate change, vaccination and, well, facts themselves is a major problem in the world. Sadly, even knowing that you are biased doesn't help. Even the scientists who study it say they struggle to recognise their own biases.

5 Your Deep Past

How evolution shaped us to dominate the world

Imagine travelling back in time through your own family history. At first you'd meet your parents, grandparents and great-grandparents. Next would come ancestors who died before you were born: 16 great-great grandparents, 32 great-great-great grandparents, and so on. Keep going further and further back and you'd soon be meeting people who lived and died in prehistory.

But they would still be human. Even after 5,000 generations your direct ancestors would be physically and mentally like you. But then gradually, ever so gradually, your family would start to change. Go back another 5,000 generations and they would be less recognisably human; shorter, stockier, heavier in the brow, sloping in the forehead.

Eventually, around 7 million years back in time, you would meet an ancestor who wasn't human at all. This is our – your – last common ancestor with chimpanzees.

Missing link

We don't know much about this creature – the fossil record for that crucial period is frustratingly sparse – but if it were alive today there's little doubt that we would consider it to be a non-human primate. It did not walk upright, was covered in fur, had a small brain, heavy jaw and no language. Yet for all its non-humanness, this is your great, great, great (repeat thousands of times) grandparent.

How did we get from there to here? In other words, what is you and your family's deep history?

The story turns out to be a series of lucky accidents. If even one of them hadn't happened, you wouldn't be here.

The first distinctly human trait to emerge was bipedalism. Even as far back as 6 million years ago, your ancestors were walking on two legs.

There are numerous ideas to explain this adaptation. Charles Darwin suggested that it freed the hands for tool making, though the earliest known tools are only 3.4 million years old. It may have allowed your ancestors to feed while carrying infants, or move about the forest canopy more easily, walking on the hind limbs and holding on with the hands. Orang-utans and other primates walk like this along branches when feeding. Maybe

Are other humans alive today?

Legends of human-like creatures in remote places, such as Bigfoot and the Yeti, have entranced people for centuries. They make for good stories, but could there be any truth in them?

Most scientists flatly reject the possibility, but a few are willing to contemplate it, pointing out that other hominin species coexisted alongside our ancestors until quite recently. *Homo floresiensis*, aka the 'hobbit', was a pint-sized hominin which lived on the Indonesian island of Flores until 18,000 years ago. Another species, the Denisovans, lived in Siberia around 30,000 years ago.

It remains a long shot, but small groups of similar cousins could still be clinging on in remote areas of Eurasia such as the Himalayas or the Caucasus.

it reduced exposure to the sun, or made it easier to scan the horizon for predators.

However it emerged, there was a second phase of evolution around 1.7 million years ago, when your ancestors had left the forests for the savannah. This is when the greatest anatomical changes took place, with shoulders pulled back, legs lengthened and a pelvis fully adapted to life on two legs.

In the meantime, something else had changed: we lost most of our body hair (see 'Why are humans so hairy?', p 68). This may also have been a response to moving out on to the sweltering savannah, which made keeping cool more of a challenge than keeping warm.

Growth spurt for the brain

A subtler but more profound change was also well under way at this point. If you were crazy enough to try to bite your own finger off, you'd find it hard work. A chimpanzee's jaws, in contrast, are so powerful it can bite off a finger in one chomp.

Wimpy jaw muscles are one of the things that differentiate us from our closest relatives. This is down to a single mutation in a gene called *MYH16*, which encodes a muscle protein. The mutation – dated to about 2 million years ago – inactivates the gene, causing our jaw muscles to be much smaller.

The loss of jaw strength may have been enabled by innovations such as cooking. It had a huge impact, clearing the way for rapid brain evolution. Other primates' jaw muscles exert a force across the whole skull, constraining its growth. But our mutation weakened this grip. A brain-growth spurt began soon after.

What drove this spurt is another matter. The environment probably presented mental challenges. Social developments would have played a part, too, as did the evolution of language (a hugely important milestone of course, but one that has proved almost impossible to pinpoint in time). A big brain is incredibly hungry: yours uses about 20 per cent of your energy at rest, compared with about 8 per cent for other primates. So early humans needed to change their diet to support it. The transition to eating meat would have helped. So would the addition of seafood about 2 million years ago, providing omega-3 fatty acids for brain building. Cooking might have helped too, by reducing the energy needed for digestion. This would have allowed ancestral humans to evolve smaller guts and devote any spare energy to brain building.

Equipped with a big brain, the world was your ancestors' oyster, and they achieved some epic migrations. *Homo erectus* made the first great trek out of Africa into Asia 1.8 million years ago. Around a million years later, the predecessors of Neanderthals found their way to Europe. And 125,000 years ago, *Homo sapiens* made an early foray into the Middle East, though it did not last.

The final push

Then, some 65,000 years ago, one group of modern humans left Africa and conquered the world – an extraordinary achievement for any species, let alone a puny, furless ape. Humans first pushed into the Arabian Peninsula and then on to the Levant. From there they spread east and west, occupying all of Eurasia and Australia by about 40,000 years ago. The last great push came around 16,000 years ago when people from the far east of Siberia crossed into the Americas.

Whatever your own personal family history, you are a direct descendant of that band of humans who lived on the savannah 65,000 years ago.

The viruses that are now part of you

So you think you're human? You (probably) look like one, and think and behave like one too. But at a genetic level it's now clear you cannot claim to be human through and through. Much of your DNA comes from viruses.

It's no small part either. Some 9 per cent of the human genome is viral in origin. Added to that are weird virus-like entities called retrotransposons, which seem to serve no purpose other than to make copies of themselves. These account for about 34 per cent. In total then, virus-like DNA makes up nearly half of your genome. All of which begs some big questions: how did it get there, has it changed our evolution and what is it doing to us now?

Living in harmony

Viruses, let's remember, are parasites. They can replicate only by infiltrating a cell from a living organism – human or otherwise – and hijacking its machinery and metabolism to create copies of themselves. We tend to think of viruses as short-term visitors – think cold viruses – or potential killers such as Ebola. But viruses can also live in relative harmony with their hosts for long periods of time.

Even if a virus starts off as a killer, over time it can reduce its level of aggression to reach a mutually agreeable pact with its host. Take the myxoma virus. Rabbits were introduced into Australia in the mid-nineteenth century and bred so prolifically that they became a major pest. In 1950, scientists released myxoma virus into the wild and within three months it had killed 99.8 per cent of rabbits in south-eastern Australia.

A few rabbits, however, carried genetic variations that enabled them to survive. Natural selection did the rest: over generations, these variations spread and today Australia's rabbits coexist with myxoma in a largely disease-free symbiosis. There is evidence that similar episodes – killer plagues followed by peaceful coexistence – have taken place during the evolution of humans, though when and by which viruses we do not know.

Infiltrating our genome

Another key aspect of the relationship between humans and viruses is a process called endogenisation, in which a virus inserts its own genetic material into its host's DNA. Probably the best-known exponent of this method is the human immunodeficiency virus. HIV is a retrovirus: its genes are formed of RNA, so in order to replicate it must first convert its RNA into DNA. Only then can it integrate that DNA into the human genome.

To be infectious, HIV must infiltrate white blood cells called lymphocytes. If it were to invade a different type of cell, its DNA would become a non-infectious endogenous retrovirus (ERV). What's more, if it infiltrated a germ cell – a sperm or egg – that ERV would be passed on to later generations.

The human genome contains thousands of ERVs from as many as 50 different virus families, suggesting that germ-line endogenisation has occurred time and again in our evolutionary past. We appear to be the survivors of a brutal yet creative succession of viral epidemics.

Evolutionary bonus

And it's not only retroviruses that are to blame. In 2010, the first genes from a different class of virus, called a bornavirus, were discovered in the genomes of a number of mammals, including humans. It seems to have wheedled its way into the germline of one of our ancestors some 40 million years ago.

Taking control

Viral DNA that has infiltrated human eggs and sperm – and so can be passed down the generations – does not only form genes. These HERVs, as they are called (see main story), can also play a role in regulating the expression of other genes. Promoters are DNA sequences that help to activate or repress the expression of genes. Of 2,000 promoters from the human genome, nearly a quarter have been shown to contain viral elements. Even an important protein such as beta-globin, one of the main constituents of the oxygen-carrying molecule haemoglobin, is partly controlled by a retroviral fragment.

For a host, the arrival of new viral genes should give it new material for evolution to work on. If a virus introduces a useful gene, we would expect natural selection to act on it, encouraging it to spread through following generations. Those conferring no benefit would be ignored, while any sequences that reduced the host's survival should be weeded out. Most ERVs will be negative or have no effect. The human genome is peppered with the ragged remains of such integrations. This may explain the origins of retrotransposons, which increasingly seem to be badly degraded leftovers of old viruses.

As for positive selection, scientists have now found many such sequences. The first discovery was of a relic of a retrovirus that inveigled its way into the primate genome just short of 40 million years ago, long before the human lineage split off. It created what's known as the W family of ERVs and the human genome contains about 650 versions of it. One of these, on chromosome 7, includes a gene called *ERVW-1*, which originally coded for a protein in the outer coat of the virus but is now essential to the working of the human placenta. The gene's expression is controlled by two other viral fragments – one from the original virus and one from a second retrovirus.

Vital role

Here, then, is proof-positive that viral DNA can play a vital role in human biology. And we have plenty of other examples. There are at least seven other viral genes active in the placenta, including *ERVFRD-1*, which is important for the organ's construction.

Human ERVs, or HERVs, also appear to be important in other biological processes. For example, genes from two classes of HERVs are expressed in large amounts in the developing embryo, though their roles are not clear. Many families of HERVs seem to be important in normal brain function. *ERVW-1* and *ERVFRD-1*, for example, are widely expressed in the adult brain.

Viruses, then, have been major contributors to human evolution and it's likely they will continue to be so, taking human development in new and uncharted directions. For us modern people, emerging viruses, such as Ebola and Zika, will continue to cause tragedies. For our descendants, however, those plagues could turn out to be vital.

Silhouettes of the dead

Hand stencils are found widely around the world in caves inhabited by our prehistoric predecessors. They are made by spitting pigment at a hand placed on a wall. This one, in the Caverne du Pont-d'Arc in southern France, is a copy of a 32,000-year-old image in the nearby Chauvet cave, which is among the earliest known depictions of the human form. Just before this time, our ancestors were thought to have undergone a creative explosion. Possibly triggered by changes in their brains, it allowed art, especially images of animals, to flourish. But why hands? Some suggest that hand stencils were ground-breaking for showing cave dwellers that they could represent any 3D object with a 2D outline. Others think hand stencils formed part of a prehistoric code – a forerunner of writing. For now, the jury is out: the intentions of these primitive artists remain inscrutable.

Credit: Jeff Pachoud/AFP/Getty

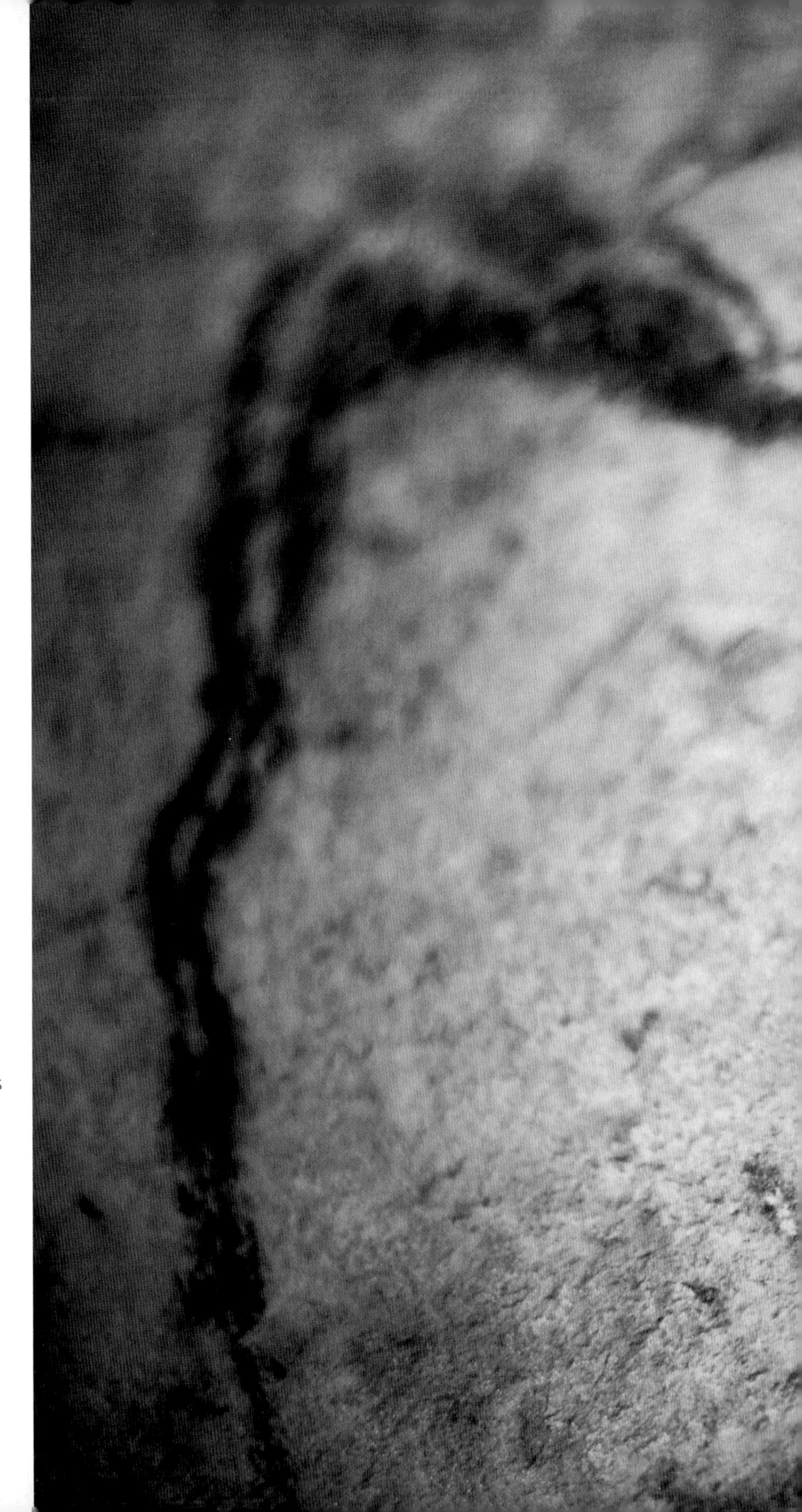

Your inner Neanderthal

Most families have a guilty secret in their history: a wayward uncle, an illegitimate child, a hushed-up adoption. The human family as a whole is no different. In fact, we are all partially the product of some illicit sexual liaisons that we might rather not think about.

We know that our closest relatives, the Neanderthals, went extinct about 40,000 years ago. Another close cousin, the Denisovans, died out around 10,000 years earlier. The assumption was that these extinctions left *Homo sapiens* standing alone. But we now know these species are not entirely gone. Traces of them are buried within all of our cells.

That's because at moments that are now (perhaps thankfully) lost in time, our direct ancestors met and had sex with these ancient human species. The genetic legacy of those remarkable encounters survives to this day. People of non-African descent carry between 2 and 4 per cent Neanderthal DNA; indigenous Melanesians get 3 to 4 per cent of theirs from Denisovans; and some hunter–gatherer groups in central Africa get a small proportion from extinct human species we haven't even identified yet – we just know they existed.

Geneticists have shown that if you combine all the ancient DNA in living humans, you could recover a sizeable chunk of the original genomes. One study suggests about 10 per cent of the Denisovan genome is still 'alive', mainly in people from Papua New Guinea. Also, about 40 per cent of the Neanderthal genome can be put together from bits living people carry and it is possible that figure will creep up as more research is done.

Red hair and freckles

We express this genetic legacy in surprising ways. It is partly responsible for the physical variation in modern humans – red hair and freckles have links with Neanderthal DNA, for instance. It also affects our health and helps us survive in extreme environments.

Like it or not, hybridisation is clearly part of our past, yet we know very little about who the hybrids were and how they coexisted with our ancestors. The genetic evidence shows that our species

Playing away from home

Our ancestors mated with other now extinct human species on at least seven occasions, and probably more. By carefully extracting DNA from ancient human fossils and analysing differences in their genomes, geneticists can make educated guesses as to when and where the matings happened. So far, they have figured out that in Europe the most recent sexual encounter between us and an extinct species was around 40,000 years ago. Some 10,000 years earlier, *Homo sapiens* were mating with Denisovans in south-east Asia. And before that, we mixed with Neanderthals on at least two occasions in what is now the Middle East. That was 60,000 years ago, not long after the great out-of-Africa migration is thought to have taken place.

produced offspring with other humans at least seven times.

If you're trying to picture how these sexual encounters happened, bear in mind that there is very little Neanderthal and Denisovan DNA on our X chromosome. This could be explained if matings were largely one-way, with male Neanderthals and Denisovans mating with female *Homo sapiens*, who then raised the hybrid children in *H. sapiens* societies. That would mean hybrid children (and their children, and their children's children) would always have an X chromosome from *Homo sapiens*, gradually diluting and diminishing any incoming X chromosomes.

Alternatively, the human X chromosome's lack of ancient DNA might be because female hybrids suffered fertility issues, and so rarely passed on their Neanderthal or Denisovan X chromosome. Or both explanations could be true.

Infertility might not have been purely down to genetics. Did the hybrids struggle to fit in socially because of their appearance? An astonishing 40,000-year-old human male jaw found in Romania could offer clues. It was found to have 9 per cent Neanderthal DNA – much more than the 2 to 4 per cent that most living non-Africans have. Geneticists say the jaw's owner must have been separated by a mere handful of generations from a Neanderthal-human hybrid. The hybrid may even have been the man's great-great grandparent. In short, the jaw is the closest we have come to glimpsing one of the hybrids themselves.

To the trained eye of some anatomists, the jaw did not look like a boring, standard-issue human ancestor. It had a curious mix of human and Neanderthal traits. The rear molars, for instance, are strikingly large compared with modern human teeth. But features that are obvious to an anatomist wouldn't have looked strange when the individual was alive. So while the subtle shapes of their bones might belie their curious genetic past, in real life hybrids probably looked the same as everyone one else, and may have been treated the same.

In fact, there is evidence that hybrids were a feature of human societies from the very beginning, which again suggests they were well integrated in human social groups. A recent study of some hunter–gatherer tribes in Africa hints that mating between early *Homo sapiens* and other ancient humans on the continent – the African equivalents of Neanderthals and Denisovans – was commonplace. This might even have given our ancestors some of the traits that allowed them to withstand the dramatic swings in the region's environment during the Pleistocene.

Mysterious humans

From statistical analysis of the hunter–gatherer genomes, we can say that this interbreeding ended only about 9,000 years ago. Sadly, 12,000-year-old skulls found in Nigeria that have a mix of archaic and modern features are the only trace that has been found so far of the other African species that lived alongside *Homo sapiens* for nearly 200,000 years. Until we find better evidence of their existence, these mysterious humans can't be given formal names – even though they may have played a pivotal role in the survival of our species.

There is an even bigger mystery. A Neanderthal toe bone found in Siberia suggests *Homo sapiens* and Neanderthals interbred 100,000 years ago – at a time when our species is thought to have been confined to Africa. The controversial conclusion is that early human pioneers must have ventured out before the main out-of-Africa migration, met their long-lost cousins, and seduced them. Or worse.

What has civilisation done to your body?

The human body is not what it used to be. Everybody knows that people have become taller thanks to better diets, and fatter as a result of overeating and less active lifestyles. But civilisation has resculpted our bodies in numerous and surprising ways over the past few thousand years.

Some of these transformations are temporary changes wrought throughout our lives that would melt away if we returned to a Stone Age environment. Others could be genetic in origin: examples of recent evolution in action. The complex interplay of nature and nurture is hard for us to disentangle, but the sheer breadth and scale of the changes show the ease with which the human body can adapt to new habitats over short timescales.

Anatomically modern humans arrived on the scene 200,000 years ago. They lived as hunter–gatherers in small nomadic groups until around 10,000 years ago, when the advent of farming led to permanent settlements and, in fits and starts, the rise of civilisation.

Rapid evolution

The idea that evolution could have been taking place in the past few thousand years goes against all received wisdom. Weren't we taught that natural selection operates over millions of years? Yet it looks like we have got this wrong. We now know that a gene giving people the ability to digest milk after infancy arose and spread with the invention of dairy herding several thousand years ago. Here's an example of civilisation resulting in physiological change. So what about physical changes?

By comparing ancient human remains, which have been preserved by accident or design, with modern-day humans we can work out just what civilisation has been doing to our bodies. Apart from becoming fatter and taller, we have also become less muscular, almost certainly because we have not been using our muscles as much. What's more, bones that no longer support large muscles can themselves become punier, so our shrinking musculature can be tracked in the fossil record.

Our bones are more spindly or 'gracile' than those of our ancestors, with the overall diameter shrinking as well as the dense outer cortex of the bone becoming thinner in cross-section. Bones are weaker too. Comparing ancient and modern thigh bones reveals a reduction in strength of 15 per cent between 2 million and 5,000 years ago. Then the trend accelerates with another 15 per cent loss over just 4,000 years.

Dwindling strength

This thinning and weakening kicked in when we began to use tools that reduced physical exertion, starting with hand axes, through to ploughs and eventually cars. Our increasingly sedentary lifestyle means our survival has come to depend less and less on our strength. How much of this process is due to genetic changes, and how much would be reversed if we lived a Stone Age lifestyle, once again is impossible to say because we don't know which genes are involved.

What we do know is that the body has an impressive capacity to respond to exertion over a single lifetime. Take professional tennis players. The humerus in their playing arm is more than 40 per cent stronger than the corresponding bone in the opposite arm. For comparison, non-athletes have only a 5 to 10 per cent difference. That's important because it suggests that we retain our ancient capacity for strength – if only we work our bodies hard enough – and stronger bones mean fewer fractures. Broken hips were less common in the past, and are vanishingly rare in archaeological specimens, even accounting for the fact that lives were shorter then.

Open structure

The spine, too, seems to be taking on a more open structure. Human remains at Pompeii, the Roman city buried when Mount Vesuvius erupted, reveal that a condition called spina bifida occulta is twice as common today as it was back then in AD 79. Around a fifth of us have the condition, in which the vertebrae in the lowest part of the spine, the sacrum, fail to form properly. In most cases there is no outward sign, although back problems may result.

Arteries lost and gained

Other physical changes are more mysterious in origin. Some of us have acquired a new blood vessel in our arms. In fact, the median artery is present in human embryos but according to textbooks it normally dwindles and vanishes around the eighth week of pregnancy. An increasing number of adults now have a median artery, up from 10 per cent at the beginning of the twentieth century to 30 per cent at the end. Over the same period, a section of the aorta lost a branch that helps supply the thyroid gland. These changes could be due to differences in the diet and lifestyle of pregnant mothers, or perhaps a relaxing of the forces of natural selection, thanks to modern medicine and welfare systems.

Even our fingerprints are changing over time. A comparison of people born before 1920 and those born later revealed differences in their patterns. Simple arches, tented arches and whorls were more common in the later group, and ulnar loops less so. Nobody knows why.

Perhaps we should be worried that civilisation has such power over our bodies. But there's a silver lining. As natural selection loses its grip on us, the human body is becoming more variable. And that could be a good thing because who knows what we'll need to adapt to in the future.

Thick between the eyes

Compare the skulls of women who lived a century ago with those alive today and you'll notice something strange. Modern women are 50 per cent more likely to have a characteristic thickening of the bone on the inside, just above the eyes. Among women in their 30s, the prevalence has nearly quadrupled from 11 to 40 per cent. Civilisation is to blame.

Smaller families, less breastfeeding, obesity, lack of exercise and the contraceptive pill all increase women's exposure to oestrogen. That's what causes the skull to thicken. It's also thought to be the main reason modern women have a one in eight chance of developing breast cancer over their lifetime.

Your royal blood

Most of us think of our family trees as including just a few close relatives, but go backwards a few generations and they grow very large indeed. Keep going back and eventually everyone is your direct ancestor. That means you are almost certainly descended from royalty

KEY

THEORETICAL NUMBER OF DIRECT ANCESTORS
Given that everybody has 2 parents, 4 grandparents and so on, the number of direct ancestors ought to double every generation

POPULATION OF ENGLAND

ACTUAL NUMBER OF DIRECT ANCESTORS
In reality, cousins interbreed, which reduces the number of direct ancestors. This 'pedigree collapse' gets bigger the further you go

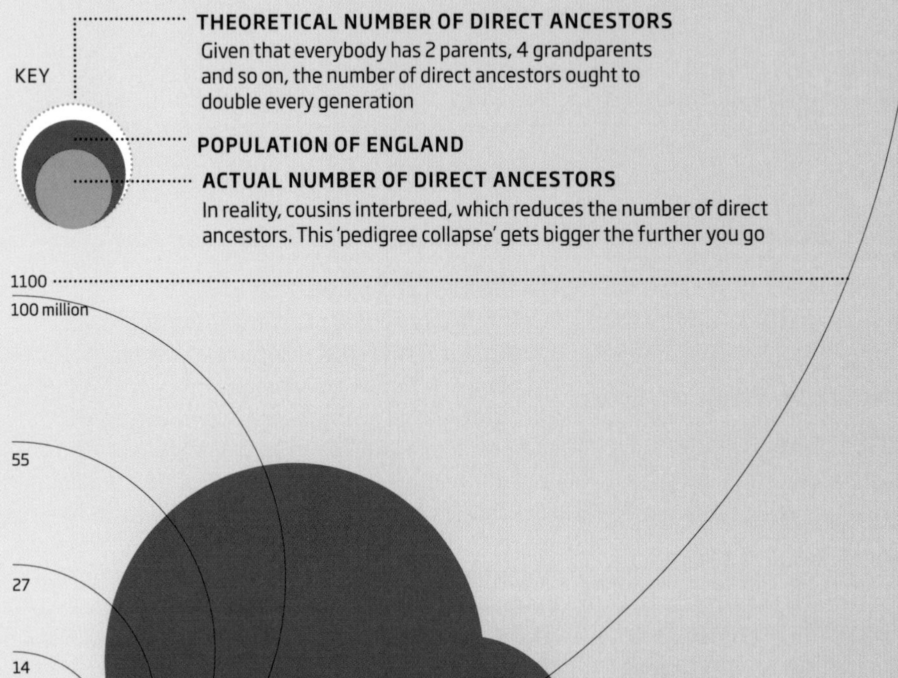

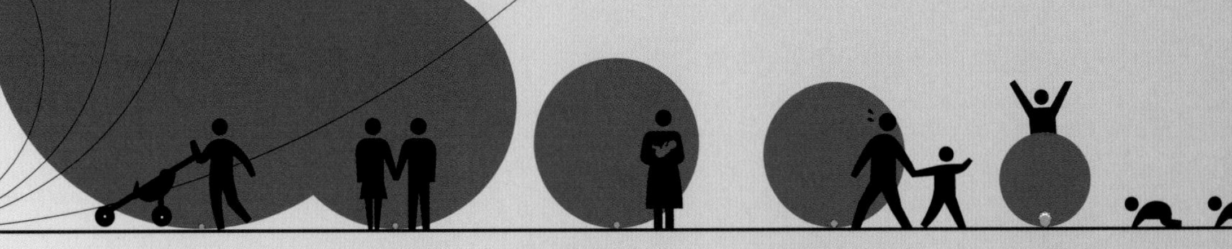

2000

Let's work backwards from here, assuming a generation time of 30 years. A 53-year-old born in 1947 starts looking into the family history

1857

Go 3 generations back and you have 8 great-grandparents, assuming no cousin marriage (which is legal but very rare)

1787

7 generations back, you have 128 great, great, great, great, great grandparents

1677

10 generations back, 1,024 people are your great, great (etc) grandparents

1527

Even this far back, cousin marriage does not have a big impact and the number of direct ancestors, 31,438, is close to the theoretical number, 32,768

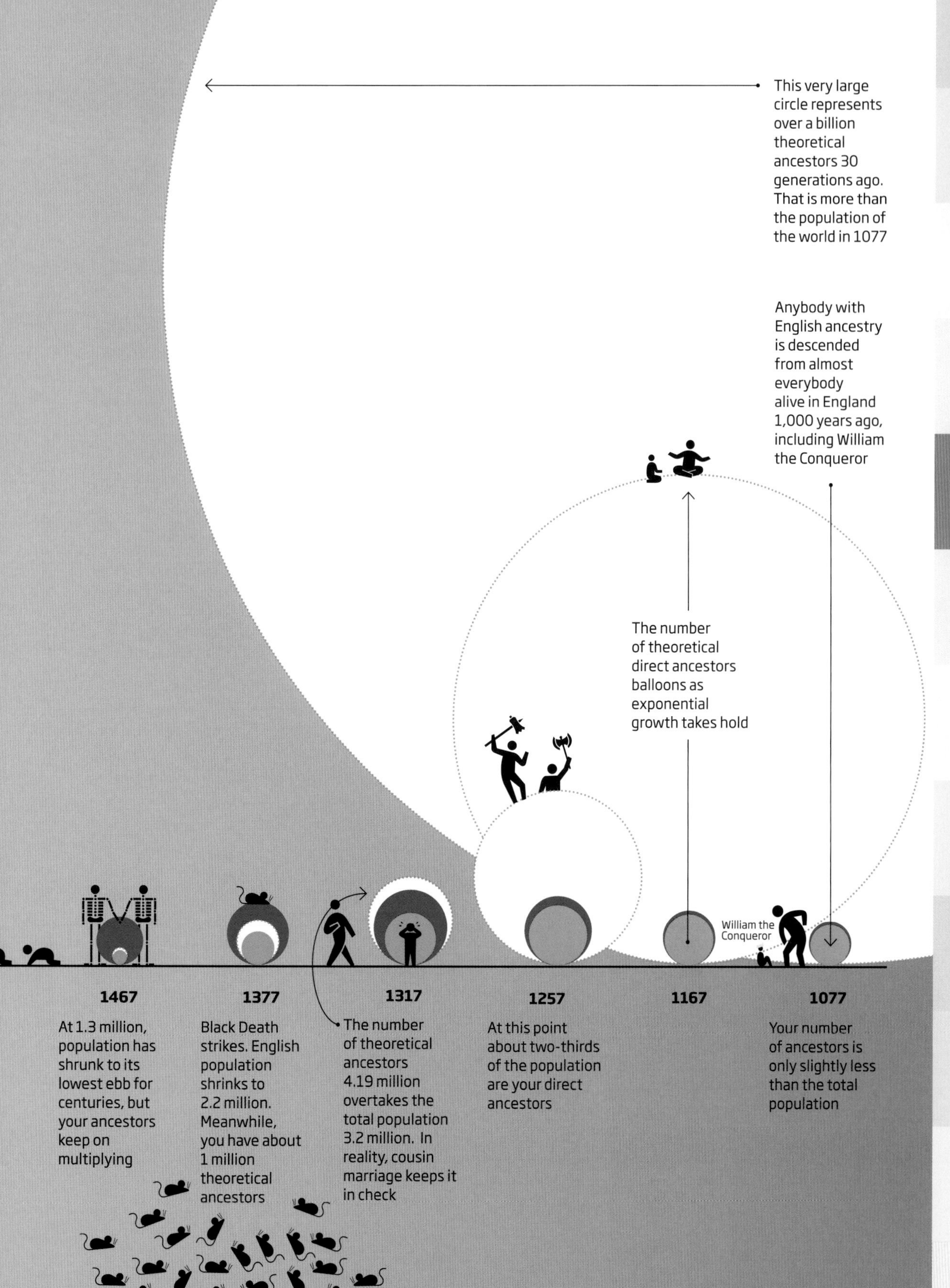
This very large circle represents over a billion theoretical ancestors 30 generations ago. That is more than the population of the world in 1077
Anybody with English ancestry is descended from almost everybody alive in England 1,000 years ago, including William the Conqueror
The number of theoretical direct ancestors balloons as exponential growth takes hold
William the Conqueror
1467
At 1.3 million, population has shrunk to its lowest ebb for centuries, but your ancestors keep on multiplying
1377
Black Death strikes. English population shrinks to 2.2 million. Meanwhile, you have about 1 million theoretical ancestors
1317
The number of theoretical ancestors 4.19 million overtakes the total population 3.2 million. In reality, cousin marriage keeps it in check
1257
At this point about two-thirds of the population are your direct ancestors
1167
1077
Your number of ancestors is only slightly less than the total population

How animals made us human

Travel almost anywhere in the world and you will see something so common that it may not even catch your attention. Wherever there are people, there are animals: animals being walked, herded, fed, watered, bathed, brushed or cuddled. Many, such as dogs, cats and sheep, are domesticated but you will also find people living alongside wild and exotic creatures such as monkeys, wolves and binturongs. Close contact with animals is not confined to one particular culture, geographic region or ethnic group. It is a universal human trait, which suggests that our desire to be with animals is deeply embedded and very ancient.

On the face of it this makes little sense. In the wild, no other mammal adopts individuals from another species; badgers do not tend hares, deer do not nurture baby squirrels, lions do not care for giraffes. There is a good reason why. Since the ultimate prize in evolution is perpetuating your genes in your offspring and their offspring, caring for an individual from another species is detrimental to your success. Every mouthful of food you give it, every bit of energy you expend keeping it warm (or cool) and safe, is food and energy that does not go to your own kin. Even if pets offer unconditional love, friendship, physical affection and joy, that cannot explain why or how our bond with other species arose in the first place. Who would bring a ferocious predator such as a wolf into their home in the hope that thousands of years later it would become a loving family pet?

To understand this fascinating puzzle, I have looked to the deep past for the origins of our intimate link with animals. What I found was a long trail, an evolutionary trajectory that I call the animal connection. What's more, this trail links to three of the most important developments in human evolution: tool-making, language and domestication. If I am correct, our affinity with other species is no mere curiosity. Instead, the animal connection is a hugely significant force that has shaped us and been instrumental in our global spread and success in the world.

The first tool-makers

The trail begins some 3.3 million years ago. That's when the first flaked stone tools appear in the archaeological record, on the shores of Lake Turkana in Kenya. Inventing stone tools is not trivial. It required the major intellectual breakthrough of understanding that the apparent properties of an object can be altered. But the prize was great. Fossilised animal bones dating from around this time have been found to bear cut marks, indicating that our ancestors were using tools to gain access to animal carcasses. Up until then, they had been largely vegetarian, upright apes. Now, instead of evolving the features that make carnivores effective hunters – such as swift locomotion, grasping claws, sharp teeth, great bodily strength and improved senses for hunting – our ancestors created their own adaptation by learning how to turn heavy, blunt stones into small, sharp items equivalent to razor blades and knives. In other words, early humans devised an evolutionary shortcut to becoming a predator.

That had many consequences. On the plus side, eating more highly nutritious meat and fat was a prerequisite to the increase in relative brain size that marks the human lineage. Since meat tends to come in larger packages than leaves, fruits or roots, meat-eaters can spend less time finding and eating food and more on activities such as learning, social interaction, observation of others and inventing more tools. On the minus side, though, preying on animals put our ancestors into

Pat Shipman is retired adjunct professor of anthropology at Pennsylvania State University, University Park. She is author of *The Animal Connection: A New Perspective on What Makes Us Human*

direct competition with the other predators that shared their ecosystem. To get the upper hand, they needed more than just tools. That's where the animal connection comes in.

Heavy competition

Three million years ago, there were 11 true carnivores in Africa. These were the ancestors of today's lions, cheetahs, leopards and three types of hyena, together with five now extinct species: a long-legged hyena, a wolf-like canid, two sabretooth cats and a 'false' sabretooth cat. All but three of these outweighed early humans, so hanging around dead animals would have been a very risky business. The new predator on the savannah would have encountered ferocious competition for prizes such as freshly killed antelope. Still, by 1.7 million years ago, two carnivore species were extinct – perhaps because of the intense competition – and our ancestor had increased enough in size that it outweighed all but four of those that remained.

Why did our lineage survive when true carnivores were going extinct? Working in social groups certainly helped, but hyenas and lions do the same. Having tools enabled early humans to remove a piece of a dead carcass quickly and take it to safety. But above all, the adaptation that made it possible for our ancestors to compete successfully with true carnivores was the ability to pay very close attention to the habits of both potential prey and potential competitors. Knowledge was power, so we acquired a deep understanding of the minds of other animals.

This had a knock-on effect. Predators require large territories in which to hunt or they soon exhaust their food supply. From the first appearance of our lineage 6 or 7 million years ago until perhaps 2 million years ago, all hominins were in Africa and nowhere else. Then early humans underwent a dramatic territorial expansion, forced by the demands of their new way of living. They spread out of Africa into Eurasia with remarkable speed, arriving as far east as Indonesia and probably China by about 1.8 million years ago. This

Animals in art

A wealth of prehistoric art appears in Europe, Asia, Africa and Australia, starting about 50,000 years ago. Prehistoric art allows us to eavesdrop on the conversations of our ancestors. And no subject was more popular than animals – their colours, shapes, habits, postures, locomotion and social habits.

This focus is even more striking when you consider what else might have been depicted. Pictures of people, social interactions and ceremonies are rare. Plants, water sources and geographic features are even scarcer, though they must have been key to survival. There are no images showing how to build shelters, make fires or create tools. Animal information mattered more than all of these.

was no intentional migration but simply a gradual expansion into new hunting grounds. First, an insight into the minds of other species had secured our success as predators, now that success had driven our expansion across Eurasia.

Crucial topic

These enormous changes in lifestyle and ecology meant that gathering, recording and sharing knowledge became more and more advantageous. So, there was an impetus to enhance communication. Nobody doubts that language was a major development in human evolution. How it arose, however, remains a mystery. I believe I am the first to propose continuity between the strong human–animal link that appeared 3.3 million years ago and the origin of language.

No words or language remain from prehistoric times, so we cannot look for them. We can, however, look for symbols – since words are essentially symbolic. And in prehistoric art we find an overwhelming predominance of animals. This was clearly the most crucial topic about which our ancestors amassed and exchanged information. The complexity and importance of that information spurred them to develop more sophisticated communication. The magical property of full language is that it is comprised of vocabulary and grammatical rules that can be combined and recombined in an infinite number of ways to convey fine shades of meaning.

As our ancestors became ever more intimately involved with animals, the third and final product of the animal connection appeared. Domestication has long been linked with farming and the keeping of stock animals, an economic and social change from hunting and gathering that started some 10,000 years ago, often called the

Creatures comfort

Our link with animals can be traced back 3.3 million years and is as crucial today as it ever was. The fundamental importance of our relationship with animals explains why interacting with them offers various physical and mental health benefits – and why the annual expenditure on items related to pets and wild animals is so enormous.

Being with animals has been instrumental in making humans human. It underpins three key developments in our evolution: tool use, language and domestication. We had best pay attention to this point as we plan for the future. If our species was born of a world rich with animals, can we continue to flourish in one where we have decimated biodiversity?

Neolithic revolution. Domestic animals are usually considered as commodities, 'walking larders', reflecting the notion that the basis of the Neolithic revolution was a drive for greater food security. But there are some fundamental flaws in this idea.

For a start, if domestication was about knowing where your next meal was coming from, then the first domesticate ought to have been a food source. It was not. The earliest known dog skull dates from 32,000 years ago. This date is controversial, since other analyses put the domestication of dogs at around 17,000 years ago, but even that means they pre-date any other domesticated animal or plant by about 5,000 years. Yet dogs are not a good choice if you want a food animal: they derive from wolves so are potentially dangerous, and worst of all, they eat meat. If the objective of domestication were to have meat to eat, you would never select an animal that eats 2 kilograms of the stuff a day.

My second objection to the idea that animals were domesticated simply for food turns on a paradox. Farming requires hungry people to set aside edible animals so as to have some to reproduce the following year. It only becomes logical not to eat all you have if you are sufficiently familiar with the species in question to know how to benefit from taking the long view. So for an animal species to become a walking larder, our ancestors must have already spent generations living intimately with it, exerting some degree of control over breeding. Who plans that far in advance for dinner?

Then there's the clincher. A domestic animal that is slaughtered for food yields little more meat than a wild one that has been hunted, yet requires more management and care. Such a system is not an improvement in food security. Instead, I believe domestication arose for a different reason, one that offsets the costs of husbandry. All domestic animals, and even semi-domesticated ones, offer a wealth of renewable resources that provide ongoing benefits as long as they are alive. They can provide power for hauling, transport and ploughing, wool or fur for warmth and weaving, milk for food, manure for fertiliser, fuel and building material, hunting assistance, protection for the family or home, and a disposal service for refuse and ordure. Domestic animals are also a mobile source of wealth, which can literally propagate itself.

Unending supply

Domestication drew upon our understanding of animals to keep them alive and well. It must have started accidentally and been a protracted reciprocal process of increasing communication that allowed us not just to tame other species but also to change their genomes by selective breeding to enhance or diminish certain traits.

The great benefit for people of this caring relationship was a continuous supply of resources that enabled them to move into previously uninhabitable parts of the world. This next milestone in human evolution would have been impossible without the sort of close observation, accumulated knowledge and improved communication skills that the animal connection started selecting for when our ancestors began hunting some 3.3 million years ago.

If I am correct, our bond with other creatures is no mere oddity – far from it. The human–animal link offers a causal connection that makes sense of three of the most important leaps in our development: the invention of stone tools, the origin of language and the domestication of animals. That makes it a sort of grand unifying theory of human evolution.

6
Possessions

Going to extremes

Survival equipment varies depending on where you are in the world. For extreme environments, the experts recommend a range of kit. One thing recommended by all of them: never go anywhere without toilet paper

Deserts

Don't use a tent, sleep outside to see the stars.

Cotton **gardening gloves** are good for shading hands and wrists

Take a **bivvy bag, sleeping bag and foil blanket and always sleep on a raised camp bed** to avoid heat-seeking scorpions

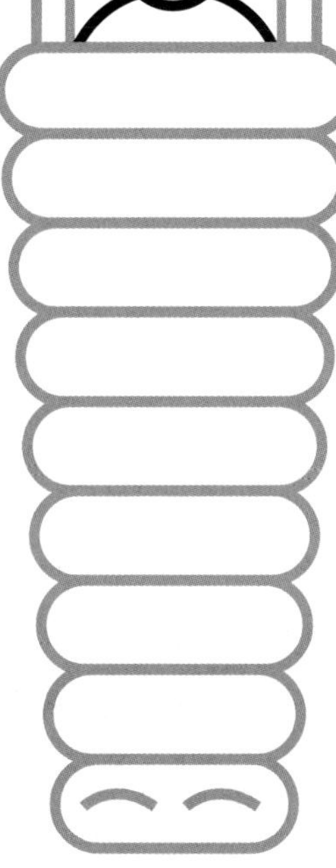

In temperatures up to 42 °C, allow 7.5 litres of **water** per person per day

Layers of clothes are best for keeping out wind, dust, dirt and occasional rain

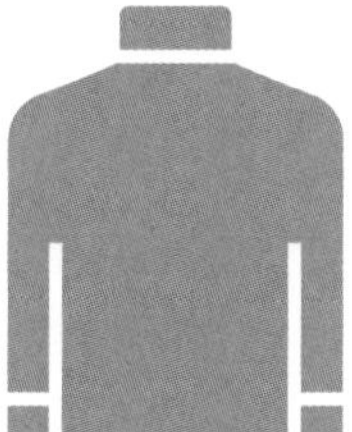

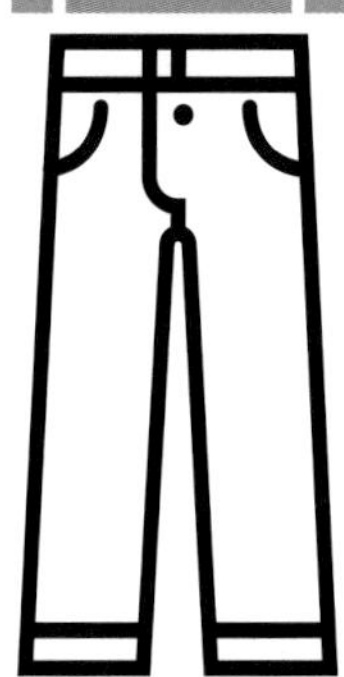

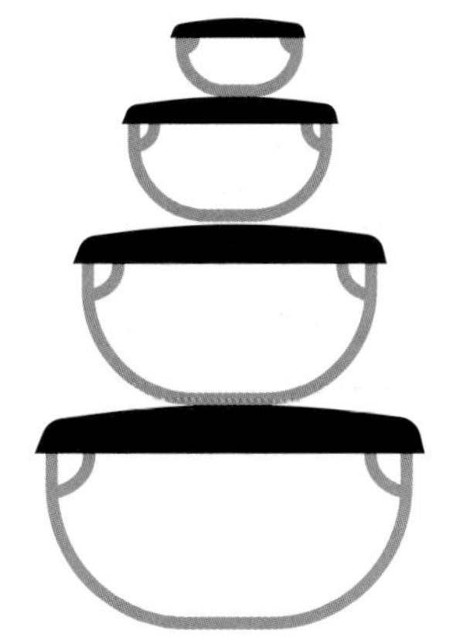

Sand gets everywhere so make sure everything is stored in **sealable containers**

Microbes in water are a real threat so take purifying tables, a microfilter or **boil water**

Sunburn is enemy number one: wear a **hat, sunglasses, sunscreen, long sleeves and long trousers**

How survival kits have changed in 5,000 years

On the social media site Instagram, thousands of people in the US post photos with the hashtag #edc, meaning 'everyday carry'. These show the tools, weapons and accoutrements that they haul around day in, day out. Men also show off the contents of their pockets through #pocketdump, whereas women tend to favour #whatsinmybag.

The core stuff is remarkably similar for both groups. The possessions we keep closest every day have a special practicality, concreteness, intimacy and symbolic importance. As the tool-making species, we are what we carry. And what we carry might offer a guide to what we really need, stripped of the clutter of overconsumption.

Tracing back in time

For an evolutionary psychologist it is natural to wonder if we can link our everyday stuff to that of our distant ancestors, for whom raw survival dictated most of their possessions.

Sadly, we don't have any prehistoric #pocketdump or #whatsinmybag images, but we do have some useful clues from Ötzi, a man who lived about 5,300 years ago, and whose ice-preserved body was found in the Italian Alps in 1991. Since then, we have learned a lot about him, from his genome and the proteins expressed in his brain to the make-up of his gut microbes and his lethal arrow wound. His possessions were also well-preserved: a diverse set of clothes, tools, weapons, fire-makers, supplies and foul-weather gear suitable for his mixed role as soldier, hunter, camper and explorer.

Much of his gear looks primitive to modern eyes. But Ötzi wasn't a distant ancestor: he had an anatomically modern brain in an anatomically modern body. In terms of timescale, we are no further from Socrates than Socrates was from Ötzi. So we should be able to find similarities between what he carried and our essentials.

Many are obvious. Ötzi's tinder fungus and flint for making fires is analogous to a lighter. His lumps of birch polypore fungus had antibiotic and anti-parasitic properties, as well as the ability to stop bleeding. It was the ancient equivalent of carrying a first-aid kit, complete with antibiotics, deworming tablets and adhesive bandages.

Leggings, boots and bag

Likewise, Ötzi's clothing and luggage make sense to us as everyday essentials. His well-worn, often-repaired goat-hide leggings are akin to a favourite pair of jeans. His deerskin shoes with bearskin

Key figure

Although not a carried possession, there is another aspect of Ötzi's life that informs today's needs. For thousands of years, his people lived in permanent settlements, usually on hilltops to protect against raids. If Ötzi was high-status, he would have lived in the equivalent of a McMansion in a gated community, with an active neighbourhood watch. Almost all #pocketdump or #whatsinmybag images include house keys. This ubiquitous portable possession unlocks warmth, shelter, security and access to the rest of our things.

Geoffrey Miller is associate professor of psychology at the University of New Mexico, Albuquerque, and author of *Must-have: The Hidden Instincts behind Everything We Buy*

soles are like rugged boots. His leather backpack is today's bag to haul around our essentials.

But it is Ötzi's weapons that get to the heart of the search for our essential possessions – namely, the ability to acquire food. His longbow was a work in progress. If he had lived long enough to finish making it, it would have been a formidable weapon, capable of killing animals up to 40 metres away. In the same vein, Ötzi's prize possession was probably his axe, which had a blade of almost pure copper. It could have chopped down trees, split firewood and defended against other humans and predatory animals. Security and warmth are core necessities for us, too.

Of course, across much of the world we no longer need weapons to acquire food and gain security, and this is where we get to the core of the things we really need today. Given modern supermarkets, hospitals, police and armies, the true analogues are the debit card, medical card, driver's licence and passport. As physical objects they are just pieces of paper and plastic, but as identity technologies, they tap into all the promises offered by vast systems of finance, medicine, security and governance.

Prehistoric pizzazz

While most of Ötzi's possessions look purely practical, it is clear that some had a bit more pizzazz. Take his stripy coat. It was made from strips of goat hide, alternating dark and light, and would have presented a striking pattern. And Ötzi's axe almost certainly carried prestige value; of his formally buried clan-mates, fewer than one in five were interred with similar axes. We start to see that even essentials can't escape that grey zone where needs mingle with wants. Like Ötzi's axe, the stuff we carry can go beyond the practical to be highly symbolic – the BMW car keys, the magnum-sized condoms, the Clinique lipstick.

Finally, our most advanced essential – the smartphone – has no analogue in Ötzi's kit. With it we can access any human knowledge, buy any good or service, and summon any form of help. We can talk with any of the 5 billion people who own a phone. We can find our location through GPS, food through Yelp, shelter through Airbnb and a mate through Match.com. If the copper axe was the most distinctive status symbol that Ötzi carried, the smartphone is ours.

Plugging in to civilisation

Clearly, at the physical level, our technologies are better, lighter and more robust than Ötzi's. Our modern boots beat Ötzi's leaky shoes. Antibiotics kill bacteria better than birch fungus.

Yet the real power of our handy essentials comes from the physical, social and informational ecosystems that they let us access. Car keys, house keys, debit cards, passports and smartphones aren't just hardware; they are the input–output devices that let our brains and bodies plug into modern civilisation. With them we can tap into vast networks of human cooperation, mutual accountability and symbolic status, on scales unimaginable to Ötzi and his peers.

So, what we need is pretty much what we carry. Next time you leave the house, grabbing your bag with your keys, phone and wallet, spare a thought for what you have with you – all the power, knowledge and vanity of an entire species compressed into a handful of objects.

What does your stuff say about you?

Take a look around your home. How much of your stuff did you acquire because it was useful or beautiful? Could you part with everything else?

The answer is almost certainly no. Our relationship with our possessions goes far beyond the practical or aesthetic. In no small measure, our belongings help to define who we are and how other people see us. They give us a sense of where we have come from, who we are and, perhaps, where we are going.

That tendency to imbue objects with rich meaning is a trait that appears early in life and develops with age. A 1977 survey of several generations of families in Chicago found that younger people value things with multiple practical uses – such as a kitchen table and chairs – whereas older people tend to prize objects that spur memories and reflection.

The endowment effect

Valuing something beyond what other people think it is worth is what psychologists call the endowment effect. It is the reason we're more likely to buy an item of clothing after we have tried it on, or a car after we've taken it for a test drive. Imagining that something is ours appears to make it more valuable to us.

The drive to acquire new things stems in part from our ability to imagine how a new thing will improve our lives, and impress others. It is this 'transformation expectation' that advertisers try so hard to exploit. And however materialistic – or not – we each view ourselves to be, we do get a shot of happiness when we buy new things.

Yet that pleasure is fleeting. Many people feel the need to top it up by buying more and more things, and are often prepared to go into debt to do so. Our consumer culture has now reached such a point that it is becoming difficult to discern where normal behaviour ends and obsession and compulsion begin.

People who seek out materialistic things to make themselves happy may be trying to fill gaps elsewhere in their lives, perhaps in their relationships. Yet the opposite does not seem to be true: the drive for more stuff does not seem to increase feelings of discontent. Loneliness, for example, tends to make people more materialistic but the inverse may not be true.

Our stuff plays an important role in bolstering our sense of identity. This effect is made all the more apparent when we are forced to let it go. This can be a difficult process, traumatic even, since it is like letting go parts of ourselves. People who have lost their homes and possessions because of a fire or other natural disaster often report a profound confusion of identity. Their pasts disappear at a stroke. Institutions such as prisons and the military try to use this effect by taking away a person's clothes and personal items and issuing them with standardised kit. The aim is to reduce their individuality in the hope that they will become ripe for reshaping.

Beyond our personal view of ourselves, we like to acquire possessions because they give us social standing and status. This is an effect that may be growing more powerful. For example, research suggests that today's twenty- to thirty-five-year-olds are more likely than previous generations to try to gain status by buying designer handbags and high-end fashion items.

Why is this happening? One suggestion is that millennials receive more money than previous generations from their parents, or because they have easier access to credit. Such trends may also explain what is called 'the fantasy gap'. Research in

When ownership goes wrong

Hoarding is a relatively new psychiatric diagnosis and only recently have scientists differentiated it from obsessive-compulsive disorder (OCD) – with the help of brain scanners. When asked whether to throw away an object, hoarders display overactivity in the anterior cingulate and insular cortex – areas that deal with importance, relevance and salience. Hoarders, it seems, worry so much about making wrong decisions that they defer making any.

Compulsive buying is not listed as a psychiatric disorder. Nobody can agree whether it shares a basis with addiction, impulse control or OCD. It is certainly distinct from hoarding, which exists in many cultures, while shopping addiction exists only in market-based economies where there is plenty to buy and income to spend.

the US shows that since the 1970s, there has been a growing divergence between older teenagers' aspirations to own expensive things and their willingness to work for them.

The power of envy

There is one other factor behind our drive for obtaining new things: 'keeping up with the Joneses'. Envy is a powerful emotion which, at a basic level, can be seen as about fairness and dignity. Is it fair that some people have plenty while others go without? And what does deprivation do to a person's self-worth? Envy is not confined to rich societies. Living standards and incomes vary in communities around the world but the influence of relative standing within those communities is ever present.

For a variety of reasons, then, the desire to possess things is very deeply ingrained in us. If we cannot easily rid ourselves of it – even if we wanted to – can we at least increase the pleasure we get from the things we buy? The answer appears to be an emphatic yes.

We know that once earnings rise to a level that can maintain a comfortable lifestyle, extra money does not improve quality of life. But that may be because people in this situation are spending wrongly. Research by psychologist Elizabeth Dunn at the University of British Columbia in Vancouver, Canada, suggests that splashing out on experiences or on other people gives more longer-term satisfaction than spending on other things. So taking the family on a great holiday may give more enduring pleasure than buying a new sound system.

Another tactic is to think about how a new purchase will really improve life. Though we have a transformation expectation, it is often nebulous and the pleasure of a purchase dissipates all too quickly. So before shelling out hard-earned money, Dunn recommends thinking through how that new thingamabob will make life better and, particularly, how it will save you that most precious of commodities – time.

BEACH TOY
PEACE
Truck
Real Billiards
BEACH TOY

Plastic fantastic
Trade may have started in prehistory to maintain food supplies in times of shortage, but it morphed into a way to obtain desirable objects and bolster status. Today, the world is home to a multi-billion-dollar trade in every conceivable object, much of it originating in places like China's Commodity City. This massive market in Yiwu city, 300 kilometres southwest of Shanghai, is the world's largest small-commodity wholesale market. Ask for a thousand widgets in the morning and you can pick them up at teatime. And it's not just gaudy plastic goods, but electronics, fashions, even car accessories. In 2017, 70 per cent of the world's Christmas decorations came from here. What would prehistoric people have made of it?

Credit: Rich Seymour/ INSTITUTE

The future of possessing

Torn. That's how many people feel about their belongings. They love their prized possessions but fret about how a world dominated by hyper-consumerism is depleting global resources and adding to mountains of waste. Fortunately, new technology offers some novel ways out of this quandary.

One obvious way to reduce the guilt over waste is to extend the lifespan of our possessions. When the lens motor on Dutchman Dave Hakkens's camera stopped working, he asked the manufacturer to repair it but received an all too familiar reply: modern electronics can't be fixed so junk it and buy a new one.

That answer inspired Hakkens, who is a designer, to create 'Phonebloks' – a mobile phone made of components that can be removed for repair or be easily replaced. He treated every constituent, from the processor to the camera and screen, as an individual module. His approach gained the support of hundreds of thousands of people in an online petition. Then, mobile phone maker Motorola announced that it too had been working on a similar approach.

Self-healing gadgets

Another route to increasing the longevity of devices is to enable them to fix themselves. When LG launched its G Flex mobile phone it added a 'self-healing' polymer coating on the back that slowly repairs minor scratches.

Similar concepts are being investigated for inside phones too. In some rechargeable batteries, the silicon anodes degrade during charging. Researchers have created a self-healing polymer coating that holds the silicon fragments together, keeping them in electrical contact and hence maintaining charge. In future, such an approach might also ensure a long life for wearable electronics embedded in our clothes.

On-demand 3D-printing also offers the possibility of extending life, especially if it enables us to make components for, say, a modular phone. Printing at home could also cut the environmental costs of delivery.

Farewell to the physical

Beyond increasing longevity, new technology promises to change our relationships with some of our favoured possessions in other ways. It used to be the case, for example, that if you wanted to

Lose the physical, go virtual

In some spheres, digital technology holds the promise of doing away entirely with physical versions of stuff. Music, photos, films and plenty more can all be stored on a single computer. Advocates of this movement admit that digital possessions, though easier to move around than physical ones, still eat up resources. They also create their own kind of clutter. Managing and organising digital files needs time and attention. And though, today, hoarding is defined in terms of physical objects, it can only be a matter of time before digital hoarders emerge.

listen to an album or read a book, you had to buy a physical object. But that has changed big time.

Back in, say, 2010 it looked as though we would soon all be downloading our music and reading ebooks. But since then, that version of the future has been replaced by a more complicated one. In the four years to 2016, figures from the British Phonographic Industry (BPI) show that digital downloads of music have fallen while sales of vinyl albums have rocketed: in 2016 they reached a level not seen since 1991. The fortunes of ebooks are also looking less certain: sales started to fall in 2015, according to data from the Association of American Publishers. Meanwhile, sales of paperbacks continue to grow.

This suggests that digital media will not supplant the older technologies but coexist with them. It seems we are choosing different formats to suit the different types of experience we want: vinyl, perhaps, for listening at home and digital for the daily run or commute. And for certain books it seems we want a physical copy sitting on our shelves at home.

Steaming streaming

There is one form of digital technology that continues to go from strength to strength – content streaming. Here, media – such as music, films and video games – are effectively rented, not owned. Think Spotify, Netflix and Steam. It is the fastest-growing revenue source for the music industry worldwide. According to the latest figures from the BPI, between 2015 and 2016 music streaming grew by two-thirds.

The digital world also promises to change our ideas about ownership through movements such as the sharing economy. Online technology is now being used to share food, rides to work and buying and selling second-hand goods. What's more, home-sharing services, such as Airbnb, continue to grow. International analysts PWC reckon that in Europe between 2013 and 2015 the sharing sector more or less tripled in value.

Emotional attachment

The big question is what impact digital technology will have on our relationships with our stuff. Some people think it could actually increase our emotional attachments to our possessions. Andy Hudson-Smith and his team at University College London attached tags bearing QR codes to items in a second-hand charity shop. People who scanned a code received information about the object's history. The project, called Shelflife, increased sales at the shop.

Hudson-Smith himself bought a tacky old bear after reading that it had been a good-luck charm for a girl who had passed her exams at school. That emotional tie got the better of him and the bear now sits on his desk as a talking point.

This demonstrates one way in which digital technology can make the ties we have with possessions more explicit. We often choose endearing objects because they remind us of places or events we have been to, or family and friends. Anything with digital connectivity or a memory can store that information for us.

This may provide a simple way for other people to find out what particular items mean to us, but will we want to miss out on telling those stories ourselves? After all, the things we care about most tend to be those with which we have the deepest emotional connections. Do we really want to convey those emotions via a digital link?

7

Friends and Relations

How does friendship work?

Most animals have acquaintances but only a few species are capable of true friendship. This select group of mammals includes the higher primates, members of the horse family, elephants, cetaceans and camelids. It is no coincidence that all these animals live in stable, bonded social groups. Group living has its benefits, but it can also be stressful and you cannot simply leave when the going gets tough – which is where friendship comes in. Friends form defensive coalitions that keep everyone else just far enough away, without driving them off completely.

Friends and onions

Friendship gives social groups a very different structure from the amorphous herds of deer or antelope. From the point of view of a member, a bonded society is made up of layers, like an onion, with your best friends at the core and successive layers filled with individuals with whom you are decreasingly intimate. Whatever the species, the core tends to consist of some five intimates, with the next layer taking the group to around 15, and the widest circle encompassing about 50 friends. Each layer provides different benefits. While intimates offer personal protection and help, you may rely on a larger friendship group for food, and the entire society for defence against predators.

It takes intelligence to live in a bonded, layered social system. Whereas a herd animal must simply know its neighbour, here you must know the structure of the whole social network of the group. This is because when you threaten me you risk upsetting my friends too, and they may come to my aid. So you must be aware of the wider social consequences of your actions.

The cognitive demands this requires are reflected in the link between the size of a species' social group and the size of its brain – or, more specifically, the frontal lobes, which are where calculations about social relationships seem to be made. This link is not straightforward, though. What matters is the complexity of individual relationships, not simply the number. So, smart monkeys, such as baboons and macaques, need bigger brains to manage groups of a given size than do less intelligent monkeys. Apes need bigger computers still.

This link between group size and brain size – sometimes called the 'social brain hypothesis' – turns out to apply not only to species but also to individuals. Neuroimaging studies of macaques and humans show that the number of friends an individual has is linked to the size of parts of their frontal lobes.

Building trust

Many species create and maintain friendships by social grooming. Grooming – or light stroking in the case of humans – triggers the release of endorphins in the brain, which makes you feel relaxed and trusting. The bigger the group, the more time an animal devotes to grooming, but the fewer individuals it grooms. This is because as group size increases and group living becomes more stressful it becomes increasingly necessary to ensure that your friends are reliable. You do this by spending more time grooming core friends.

Since the quality of a relationship depends on the time invested in it, and there are only so many hours in a day, this sets an upper limit on the number of friends an animal can have, and hence the size of the social group. Try to groom too many individuals and you spread your time too thinly, the quality of your friendships is poorer and social

Robin Dunbar is professor of evolutionary psychology at the University of Oxford, UK, and author of *How Many Friends Does One Person Need?*

groups keep breaking up. In monkeys and apes, this sets an upper limit on average social group size of about 50, which is what you find with baboons and chimps.

But humans are different. Over the past 2 million years, we have evolved ever-larger social groups. Based on the social brain hypothesis, I have calculated our social group size at around 150. This has come to be known as 'Dunbar's number' and turns out to be both a common community size in human social organisations and the typical size of personal social networks. How could humans and their ancestors have sustained groups which exceed the number that can be bonded by grooming?

It seems we have exploited three extra behaviours that are good at triggering the release of endorphins but can be done in groups, allowing several individuals to be 'groomed' at the same time. First came laughter, which we share with the great apes. Laughter typically involves a group of three people, making it more efficient than grooming one individual at a time as a bonding mechanism. Next, perhaps 500,000 years ago, we added singing and dancing, which increased the grooming group still further. Finally, language gave greater control over laughter – through jokes – and song and dance. Ultimately, it allowed rituals to be associated with religion, and this made super-groups possible.

Prone to decay

Even though we can feel a bond with a super-group consisting of thousands, most of us have no more than 150 in our personal social network. About half of these are family, and tend to stay constant throughout life. But non-kin friendships are susceptible to decay if we do not invest in them. Failure to spend time with a friend for a year reduces the quality of that friendship by about one-third.

Each of us has a characteristic pattern in the way we distribute our social capital, whether measured as time spent contacting friends or emotional closeness to them. Our best friend, for example, gets the same amount of time no matter who they happen to be. This pattern is rather like a personal social signature, and remains fixed even when our friends change.

I know what you're thinking

Many aspects of cognition are needed for making complex social decisions, but one that seems to be especially important is 'mentalising' – the ability to understand another's state of mind from their overt behaviour (see 'Whose mind will you read today?', p 98). 'I believe that you suppose that I wonder whether you think that I intend to . . .' represents five mind states, and is what human adults can typically manage. The size of key regions of your prefrontal cortex determines your mentalising skills, which in turn determine the number of friends you have.

Friends with benefits for mind and body

Friends. They have a positive impact on our health, wealth and mental well-being. Social isolation, on the other hand, creates feelings akin to physical pain and leaves us stressed and susceptible to illness. In fact, our bodies react to a lack of friends as if a crucial biological need is going unmet. This is not surprising. For us humans, friends are not an optional extra – we have evolved to rely on them.

But friendship comes at a cost; time spent socialising could be used in other survival activities such as preparing food, having sex and sleeping. Besides, just because something is good for us doesn't mean we will do it. That's why evolution has equipped us with the desire to make friends and spend time with them. Like sex, eating or anything else a species needs to survive, friendship is driven by a system of reinforcement and reward. Being friendly is linked with the release of neurotransmitters in the brain and biochemicals in the body that make us feel good.

The cuddle chemical

Understanding what motivates friendship begins in a seemingly unlikely place – with lactation. As a baby suckles, a neuropeptide called oxytocin is released from the mother's pituitary gland. This causes muscles in the breast to contract so milk flows, but it also reduces anxiety, blood pressure and heart rate. For mothers and babies, the relaxed feeling produced by oxytocin encourages suckling and helps create a strong, loving bond.

This occurs in all mammals, but in humans and the few other species that make friends, the system has been co-opted and expanded by evolution. Oxytocin has become associated with relationships beyond the mother–child bond. You release it in response to types of positive physical contact with another person, including hugs, light touches and massage. The resulting pleasant feeling is your reward and encourages you to see that person again.

Oxytocin works in other ways too. It promotes prosocial decisions, increases feelings of trust and encourages generosity.

Nor is it the only chemical driver of friendship. Another key player is the group of opioid chemicals called endorphins. These are released in response to mild pain, such as exercise, and act as neurotransmitters in the brain to create a feeling of well-being. All vertebrates produce endorphins, so they must have evolved early, but like oxytocin they have come to play a role in motivating friendship – and not only by making physical contact feel good.

Robin Dunbar and colleagues at the University of Oxford asked people to row a boat, either alone or in pairs, and measured their endorphin levels before and after. Their findings were striking. Despite exerting the same physical effort, people who rowed their boat as a pair released more endorphins than those who rowed alone. A major component of true friendship is behavioural synchrony – friends must be in the same place at the same time to establish and maintain a relationship. Endorphins seem to promote friendship by making synchrony feel good.

Gathering gossip

To select, acquire and maintain friends we need to gather social information. Again, this is something we enjoy. Even before babies can speak, they prefer looking at faces over other visual stimuli. We find social information intrinsically rewarding because it triggers reward-related areas of the brain. Show people in an MRI scanner pictures from their Facebook accounts, and the nucleus accumbens

Lauren Brent is lecturer in animal behaviour at the University of Exeter, UK

lights up – a brain region associated with drug addiction. Those with the greatest response are the most frequent social media users. Some people, of course, are friendlier than others. They may simply be better at making friends, but such MRI results hint that they may also do so because it gives them a bigger kick.

Popularity factors

Friendlier people are more sociable, in part, because their genes make them that way. James Fowler at the University of California, San Diego, and Nicholas Christakis at Yale University compared the social networks of identical twins, who share all their genes, and paternal twins, who share 50 per cent on average. They found that genetic factors accounted for 46 per cent of differences in how popular among their peers individuals were.

Of the many people we encounter in life, how do we pick a select few to be friends? The answer, at first, seems simple – we choose people similar to ourselves, whether they are the same age, gender or profession. But it turns out that this tendency for like to associate with like also has a basis in our genes. Fowler and Christakis found that people are as genetically similar to their non-kin friends as they would be to fourth cousins. One of the mysteries of friendship has been why we cooperate so readily with complete strangers. In evolutionary terms, we should cooperate with kin because we succeed by proxy if they pass to future generations more of the genes they share with us. But if friends are more genetically similar than we would expect by chance, perhaps we should think of them less as strangers and more as 'facultative relatives'.

So, your genome may help determine not only how friendly you are, but also who you choose for your friends. No one knows how we recognise people who are genetically similar. It could be similarities in facial features, voice, gestures, smell or even personality. Whatever attracts us, one thing is certain: befriending these people will be rewarding. The one thing we know about friendship is that it feels good.

Going solo

Lonely people have elevated levels of the stress hormone cortisol. Chronic stress damages health, which probably explains why social isolation increases the risk of cardiovascular diseases and susceptibility to infection. But stress can be useful: it acts as a warning that homeostasis – the body's maintenance of stable internal conditions – has been disrupted. So stress prompts us to behave in ways that restore homeostasis, including resting when tired and seeking shade when hot. Perhaps it also motivates us to seek out social contact when we are lonely. The fact that we produce less cortisol in a stressful situation if we have a friend with us suggests that friends either help us to restore homeostasis or prevent its disruption in the first place.

40
40
TNT
Baseball
Izabelle

Seeing double
Welcome to Twins Day! Every year, thousands of pairs of twins from around the world descend on Twinsburg, Ohio, to celebrate 'twinness'. It's an unwritten rule that identical twins dress identically. And wherever twins gather, medical researchers follow. Twin studies have become a powerful tool for teasing apart the contributions of nature and nurture to investigate what makes us who we are. Some of the most intriguing studies follow identical twins separated at birth who share the same genes but have very different upbringings. Such studies have been used to explore diseases and traits including breast cancer, musical pitch perception and sense of humour. This trait, measured by reactions to Gary Larson cartoons, seems to be totally outside genetic control.

Credit: Susana Raab/ INSTITUTE

The unseen impacts of brothers and sisters

You can choose your friends, but you can't choose your family. How unfortunate, then, that families – especially older siblings – can have a profound effect on the course of your life.

You've probably heard popular ideas of firstborn over-achievers, neglected middle children and spoilt younger ones. And it turns out there's some truth to these stereotypes. A growing body of research reveals that your position in the pecking order really can affect your life in these ways, and others too.

Smartest first

The firstborn overachiever label may reflect the fact that oldest children tend to be more intelligent than their siblings. Eldest brothers have about a 2.3-point advantage over second brothers in IQ score – a trend that continues down the birth order. Curiously, in families where an elder brother dies, second brothers score higher in IQ tests than expected from their original place in the birth order. This suggests that what counts may not be order of birth per se but social position.

Firstborns also tend to be taller than their siblings, by about 2.5 centimetres after correcting for socioeconomic status, ethnicity and parental height. Height is a desirable physical attribute and may also contribute to overachiever status. But firstborns also have to contend with being fatter. One explanation for this may be that blood vessels in the placenta of first pregnancies do not develop fully, so the first child is relatively undernourished and lighter at birth than subsequent children. People whose metabolism develops in the womb during a period of scarcity and who later in life are exposed to plenty are thought to be at higher risk of becoming obese later in life.

Other research points in a similar direction. Even while they are children, firstborns are less responsive to insulin than their siblings and tend to have significantly higher blood pressure. Such traits are likely to put them at higher risk of developing type-2 diabetes, heart disease, hypertension and stroke.

There is also a link between birth order and allergies. Children of four and five years who have older siblings are much less likely to be rushed to hospital with allergic asthma than children without older siblings. This effect is often explained by the hygiene hypothesis, which argues that exposure to a broad range of microbes in early life helps children to develop robust immune defences. An older child who is likely to distribute novel viruses, fungi and bacteria to his or her siblings should strengthen that effect, helping to boost their immune systems and make them less prone to allergies.

Immune interplay

An alternative idea stems from immune responses during pregnancy. A mother's body must reduce its immune response so it doesn't reject the growing fetus. The fetus must do the same so as not to damage its mother. One suggestion is that the mother tempers her immune response more efficiently with each successive pregnancy, so later fetuses develop fewer antibodies to her. Consequently, their immune systems are less likely to overreact to harmless substances, such as pollen, once they are born.

Birth order also affects family relationships. First-and last-born children tend to have the closest relationships with their parents, while middle children develop stronger ties outside the family. Scientists speculate that middle children tend to receive less parental attention and so look elsewhere for friendship. In other words,

A matter of life or death

Perhaps the most surprising influence of birth order is that it can have life-or-death consequences. The more older siblings a person has, the more likely they are to commit suicide. The effect is larger in women, even though their suicide rate is much lower than that of men. Why birth order should have this influence is not clear, but one idea is that firstborns have their parents' full attention for the first few years of life, which helps them to develop stable personalities that are more resilient to stress later in life. Later siblings may not be so lucky.

middle-child syndrome and the mollycoddled baby are real.

Daring youngsters

Personality, too, is influenced by birth order. Animal studies show that when zebra finches reach adulthood the youngest birds from the clutch have a greater appetite for exploring novel surroundings than their older siblings. The same may be true of humans: younger siblings have been found to be more likely to take part in dangerous sports. Also, among brothers who play professional baseball, younger siblings tend to be more likely to risk stealing a base.

There is evidence that last-born children are more adventurous politically too. Research in the Netherlands shows that 36 per cent of men and women in public office are firstborns, while 19 per cent are last-born. The oldest and youngest children in a family each make up about a quarter of the Dutch population, so firstborns are over-represented in mainstream politics, while last-borns are under-represented. When it comes to leadership that challenges the status quo, however, the tables turn: later-borns are more likely to take up revolutionary causes.

Sexual preference

Birth order may even influence our sexual preferences: a number of studies have shown that the more older brothers a man has, the more likely he is to be gay.

Why this should happen is still uncertain, but a leading candidate is – once again – the mother's immune response to her fetus. In this case, though, whatever the effect is seems to become stronger with each pregnancy. Suspicion has fallen on maternal antibodies that bind to proteins on the surface of male fetal brain cells in the anterior hypothalamus, an area associated with sexual orientation. One idea under scrutiny is that if these antibodies change the role of the proteins in typical sexual differentiation, that might lead some later-born males to be attracted to men.

Birth order clearly has implications for many aspects of our lives, though how these compare with other environmental and genetic influences we can't say. But it might be important to find out. Falling fertility rates around the globe mean a growing percentage of people are firstborns. We may be heading for a world where fat, stroke-prone, conservatives are in the majority.

How touch affects your self-image and social success

We are a hands-on species. We use touch to explore the world and turn it to our advantage. We also touch one another, and between relatives and friends we are now finding that a reassuring pat on the back or soft caress may be far more important to us than we imagined.

Research is revealing that touch gives the world an emotional context, helping us to build friendships and trust. What's more, it guides the development of brain regions that govern social behaviour and even helps to create our sense of self. It seems the touch of other people is no mere sentimental indulgence.

Cinderella sense

When it comes to the senses, seeing and hearing receive far more attention than touching. Yet the skin is our largest sense organ. On an average-sized man it weighs up to 6 kilograms and is packed with sensory receptors.

The nerves that carry signals from these receptors to the brain come in different forms. Type A fibres are heavy-duty wires that waste no time in conveying urgent news of pain, pressure, vibration and temperature changes. At the other end of the scale, type C fibres are thin and carry information slowly – about a second from ankle to brain. They were thought to respond to less immediate aspects of pain, such as aches and throbs.

Then, in the late 1990s, scientists identified another form of C fibres in humans, called C-tactile or CT fibres, that are activated by soft caresses. While most touch receptors are concentrated in places like the lips and fingertips, the receptors for CT fibres are found only in hairy skin, so are found mostly atop the head, upper torso, arms and thighs.

In the brain, nerve fibres convey information about touch to a receiving centre called the somatosensory cortex. But CT fibres also plug into the insular cortex, which is involved in emotions, and has links to a network of regions involved in thinking about other people and their intentions. These regions respond to a variety of other social cues, such as facial expressions, raising the idea that touch may be another way to communicate social intentions.

The CT fibre system seems to be primed to respond to low-force, low-speed strokes of between 3 and 5 centimetres a second. It also responds more to warm temperatures than cold. In other words, it would appear to be tuned precisely to the kind of touching – called social or emotional

Distant touch

The rise of the web, email and social media raises the question of whether we are missing out on touch in our everyday communication. Inventors are working to make sure this does not happen. The Hug Shirt is a sweater laden with sensors with which you can digitally 'record' a cuddle by hugging yourself. A file of that hug is then sent to a loved one wearing another shirt, which recreates the embrace – its strength and duration. Similar pairs of devices are using haptic technology to convey strokes and squeezes, temperature changes and vibrations over the internet.

touching – that takes place between a parent and baby, or between lovers.

For evolutionary psychologists such as Robin Dunbar at the University of Oxford, this is no surprise. He sees gentle stroking in humans as akin to grooming in other social animals, such as monkeys, where it serves to reinforce bonds between individuals (see 'How does friendship work?', p 144). Caressing and grooming trigger the release of feel-good neurotransmitters, called endorphins, which encourage individuals to spend time together and so develop trust.

Bodily boundaries

If emotional touching helps us in our everyday interactions, it may also be important in developing our sense of self. The rubber-hand illusion is an experiment in which a person's hand is put behind a screen and stroked gently at exactly the same time as a rubber hand which is in full view. After a time, many people perceive the rubber hand as the real thing: their sense of bodily self extends to include the rubber copy (see 'Where do I begin and end?', p 90).

This experiment works best if the stroking takes place at a speed known to excite CT fibres, leading some researchers to conclude that emotional touching has a critical role in teaching us the boundaries of our own bodies – what physically belongs to us and what does not. Babies are not born with a complete sense of body and self so, the reasoning goes, parental caresses are important in helping to build them.

Making the right links

Our social skills may also be early beneficiaries of touch; very early beneficiaries. Touch is the first sense to develop in the womb, and is active from about eight weeks after conception. There is plenty for a fetus to feel, through sucking a thumb, grabbing the umbilical cord or bumping up against the mother's abdomen.

Emotional touching may also be important here. Amniotic fluid swirling across the downy hairs that cover a fetus may stimulate the brain to develop, creating neural links with the insular cortex and the network of other regions that are used to interpret social cues.

These links continue to develop after the child is born. Similar stimulation is needed for other sensory systems to develop. Animal experiments show that if an eye is covered soon after birth, the parts of the brain that would normally process images from that eye will never develop, even if the cover is later removed.

Clues to autism

The emerging view of the importance of touch, especially to babies, may yield new insights into autism. Studies suggest that children with autism process touch differently. While they are perfectly aware that someone is touching them, they seem to miss the social significance of the contact.

Putting all the evidence together, some researchers contend that tactile interaction, first between mother and fetus, and later through the caresses and cuddles given by parent to child, lays the groundwork for a well-adjusted social brain. If this proves correct, then emotional touching is not just nice to have, it's essential to our development.

It's a small world

If you have played 'six degrees of Kevin Bacon' you're familiar with the idea of a small-world network – that everybody is connected to everybody else via six or fewer mutual acquaintances. The ultimate version of this is the Erdős-Bacon-Sabbath game which connects people to Bacon, the prolific Hungarian mathematician Paul Erdős and the heavy metal band Black Sabbath.

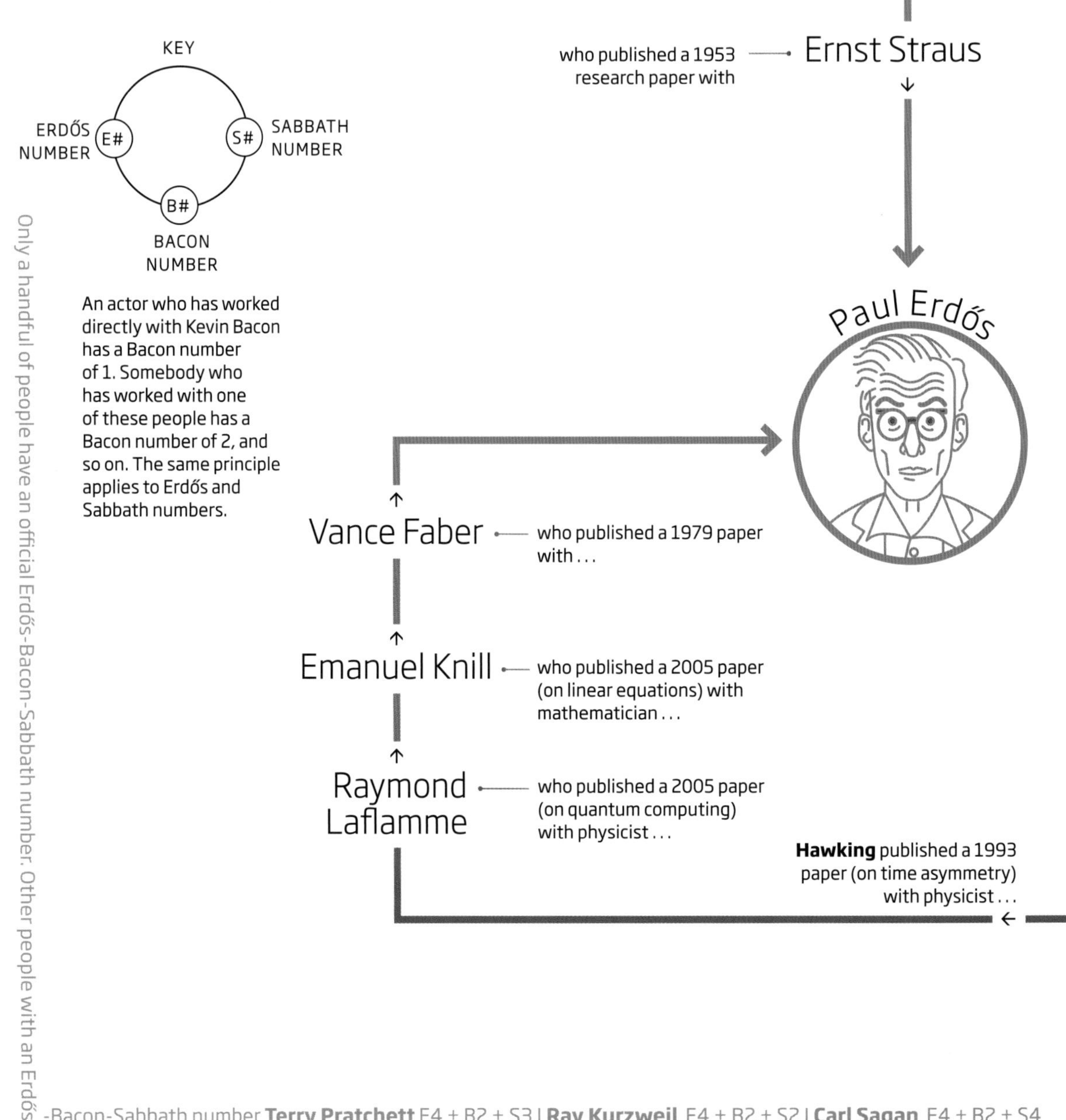

Only a handful of people have an official Erdős-Bacon-Sabbath number. Other people with an Erdős-Bacon-Sabbath number **Terry Pratchett** E4 + B2 + S3 | **Ray Kurzweil** E4 + B2 + S2 | **Carl Sagan** E4 + B2 + S4

You don't have to be a scientist or artist to be connected like this. Experiments have shown that randomly selected people are usually connected to other randomly selected people via 6 to 7 social connections

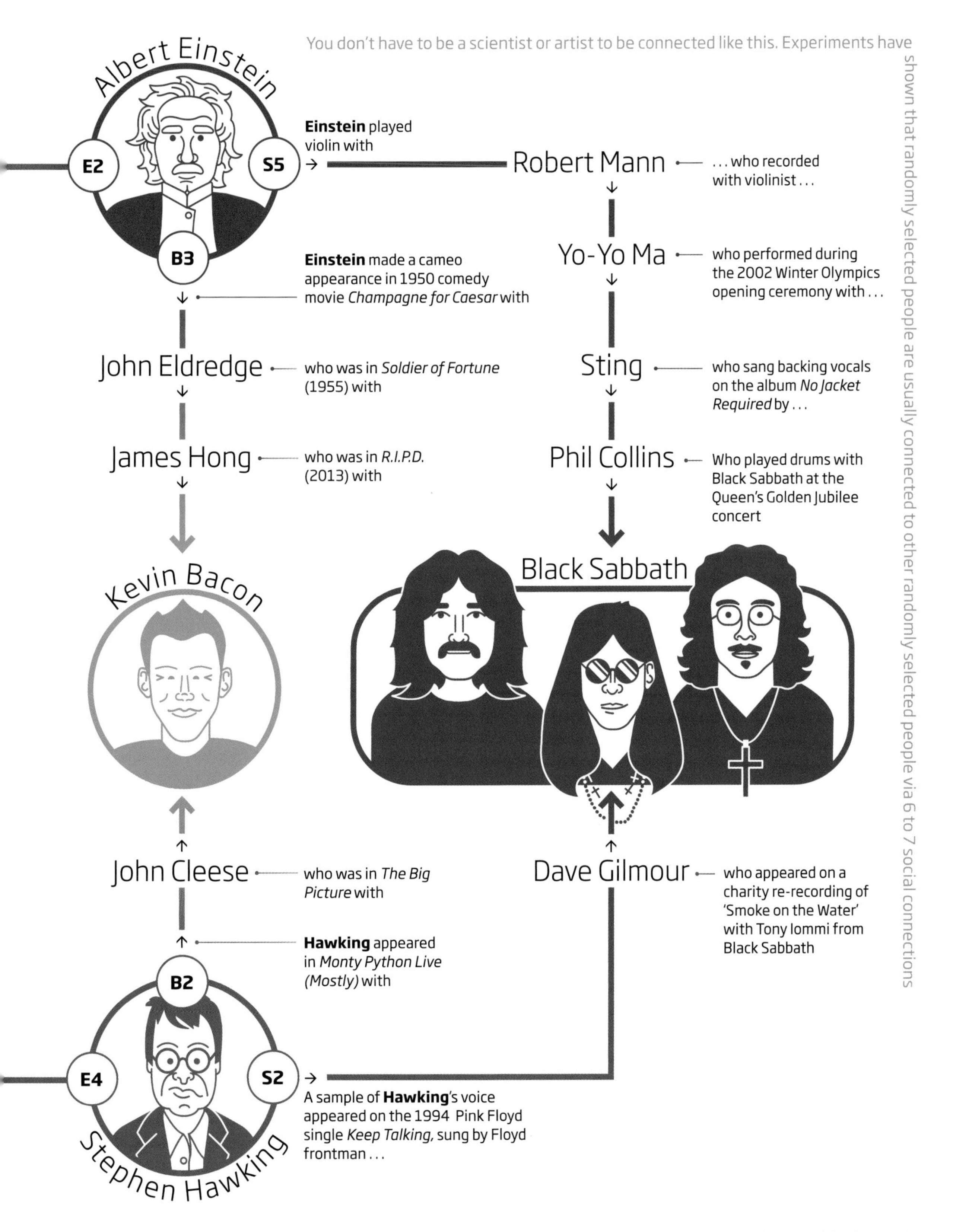

Condoleezza Rice E6 + B3 + S4 | **Noam Chomsky** E4 + B3 + S4 | See more at http://erdosbaconsabbath.com/

Why you should sniff a potential partner

Body odour – ewww! That's most people's reaction. Many of us go to great lengths to disguise or expunge our personal smells with deodorants and enigmatically named perfumes that promise to make us more appealing. Yet growing evidence suggests that components of body odour play a crucial role in who finds us attractive and who we find alluring.

Led by the nose

We tend to think of ourselves as mostly visual creatures. But it's becoming clear that our sense of smell directs our lives more than we have ever realised. To start with we have more apocrine sweat glands around the hair follicles of our armpits and groins than any of our nearest evolutionary cousins. Bacteria feast on their secretions, and in doing so release chemicals that add to our personal perfume. Not for nothing has *Homo sapiens* sometimes been called the scented ape.

What's more, our sense of smell is more sensitive than we previously thought. We are good at sniffing out information about other people – especially members of the opposite sex.

Much of this learning has come from variants of tests in which, for example, a group of male subjects is asked to sleep alone in a T-shirt and refrain from smoking, consuming pungent food and alcohol or using perfumes. A group of women – and sometimes men – is then asked to sniff each T-shirt and score it for certain properties, and perhaps rate other information of the wearer, such as a photo.

In one such study, subjects also filled in personality tests while the sniffers rated the T-shirts for various personality traits. It seems we are especially good at judging levels of extroversion, neuroticism and dominance. The first two of these tend to be emotional traits that can affect the rate at which we sweat, and hence our smell. Dominance is associated with higher levels of certain hormones that break down into molecules that could influence odour.

Such studies show that women tend to prefer men who smell dominant. This is especially true when women are at the most fertile stage of their menstrual cycle. Women can also somehow smell a man's body shape, tending to prefer the smell of men with more symmetrical bodies. High body symmetry is thought to indicate an ability to withstand infections and other environmental stresses: a sign of genetic quality.

Giving the game away

In turn, men tend to prefer the scent of women who are ovulating. Unlike the females of many animal species, women show few external signs of ovulation, but their scent does seem to give the game away. And men like it, even if only subconsciously. Conversely, they find women who are menstruating less appealing.

Here, then, may be targets for the perfume industry. The scent of fertility might make women more attractive, and that of a symmetrical body could do the same for men. Unfortunately, we don't yet know the relevant chemicals in these two cases. But we may be closer to finding the scent of dominance.

Dominant male mice produce large amounts of androstenes, which are breakdown products of male steroid hormones, such as testosterone. These compounds also seem to work in humans: dab androstenes on the top lip of women and they tend to rate men as more attractive. There is a snag, however. Not all women fall for the odour of androstenes: some cannot smell them at all and others find them unpleasant.

This finding, that no single scent is likely to work for everyone, is also inherent in another set of chemicals that influence our personal odour – the major histocompatibility complex. The MHC is a set of molecules that binds to foreign bodies and presents them to immune cells for disposal.

Smell the difference

Research shows that we favour the smell of people of the opposite sex whose MHC genes are very different from our own. This preference may have originated to avoid inbreeding when humans lived in small groups, or as a way to give children the widest variety of MHC genes, and hence the broadest immune protection.

But a single MHC-based odour will inevitably put some people off. So, a more nuanced approach is needed. It turns out that MHC genes influence our preferences not only for body odour, but for other smells too, including perfumes. So rather than design perfumes for people of different sexes, ages and incomes, it may pay perfumers to produce scents that accord with different MHC preferences.

But even this approach has a drawback. As individuals, we are good at subconsciously choosing scents that complement our own body odour. But we are pants at picking perfumes for others. Choosing a scent for a mate can actually make them smell worse, even if it's a highly rated brand. Those in the know agree that it's best to let our lovers make their own choices.

The confounding factor here is the interaction between the wearer's odour and the bottled scent. All manner of factors can change how a scent works on a person's skin, not least their diet. Perfume houses already recognise this by making perfumes to meet different regional and cultural preferences around the globe.

Despite the difficulties, our improved understanding of body odour and its components is bound to open new avenues for perfumers. One scent that works for all looks unlikely, but the idea of personalised perfumes, akin to personalised medicine, is certainly not to be sniffed at.

Scent of strong immunity

The link between our immune system and body odour is still a mystery. A set of molecules called the major histocompatibility complex helps to see off bacteria, viruses and other invaders (see main story). And it changes our body odour. But how? One possibility is that MHC molecules expressed by skin cells affect which bacteria can live on them, and consequently the odours produced. Alternatively, the smells may come directly from peptide ligands, the part of MHC molecules that binds to invaders. Perfume houses have already started patenting these peptides for creating scents.

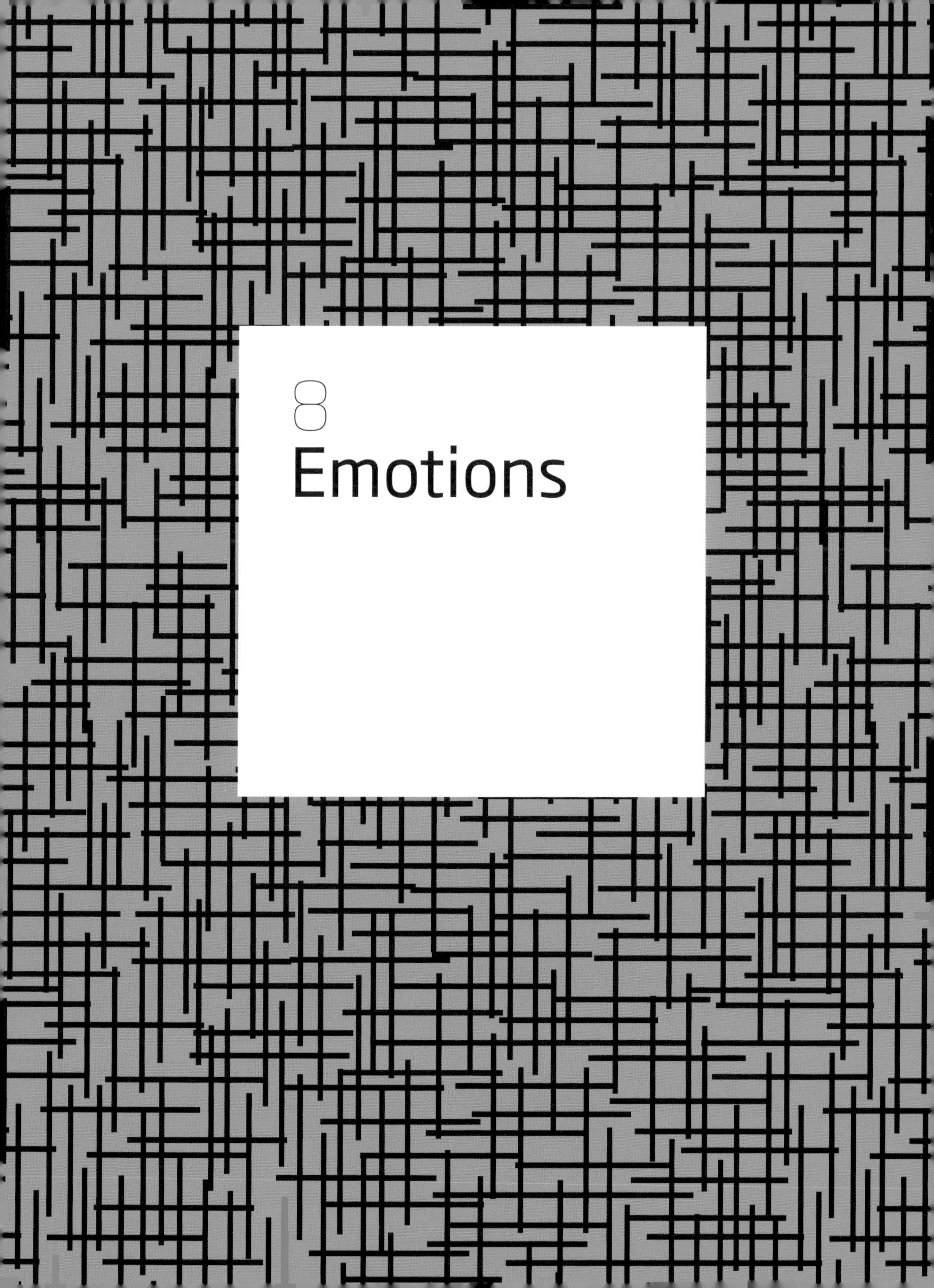

8
Emotions

Welcome to a world of emotions

How do you feel right now? Even if you're not highly emotional, you probably feel *something*. Emotions are our constant companions in life and exert a powerful influence on how we think and behave. Life would be immeasurably diminished without them. But what are they for?

Charles Darwin was one of the first to seriously contemplate the question. In the late 1860s he liked to show visitors to his house photographs of people whose facial muscles were being shocked by electrodes, contorting them into ghoulish expressions. Darwin was fascinated by how a twitch of the mouth here or furrow of the brow there could conjure up an emotion – fear, say, or surprise. He wanted to know whether his guests perceived the same emotions. They usually did.

What are they good for?

In 1872 he published *The Expression of the Emotions in Man and Animals*, in which he made the case that emotional expressions were innate and universal. But Darwin was at a loss to explain what purposes those expressions served – he thought they were probably 'not of the least use'. It is now widely accepted that Darwin was wrong.

First of all, the six basic emotions – happiness, sadness, fear, anger, surprise and disgust – play a key role in survival. They evolved as mechanisms to instigate behaviours that improve our chances of (a) staying alive and (b) procreating.

Fear, for instance, is the brain and body's way of saying 'Get the hell out of there and don't go back!' Fearless animals would have been less likely to survive.

Facial expressions play an important part. When people make a classic fear expression, they increase their field of vision, make more rapid eye movements, and open their airways – all of which allows them to better see danger and respond.

Disgust is another emotion with an obvious survival function. The feeling of revulsion repels us from harmful things such as faeces, vomit or rotting food. The scrunched-up facial expression constricts the airways to stop infectious agents entering.

For the other basic emotions, physiological function is less obvious. But in general, emotions reward us for behaviour that enhances our chances of surviving and breeding or punishes us for doing dangerous things.

Show your feelings

Emotional facial expressions have another role too: they transmit your inner state to others in order to manipulate their behaviour to your advantage. An angry face signals a threat that may cause an opponent to back down. A sad one elicits sympathy or signals a willingness to appease. And when you're smiling, the whole world smiles with you.

The big six are not the be-all and end-all of human emotions, however, as a quick audit of your own emotional state will reveal. It's unlikely that you're feeling happiness, sadness, fear, anger, surprise or disgust, but something more subtle. Curiosity, perhaps, or anticipation (hopefully not boredom or disappointment).

Modern emotions

The list of these secondary emotions is long, but they are perhaps more relevant to everyday life. Our ancestors may have had daily need of fear to flee predators, anger to conquer foes and disgust to avoid diseases, but we live in a more subtle world in which other emotions have come to the fore.

Consider curiosity. This is the feeling of arousal when a novel stimulus inspires the need to explore

Feeling awumbuk?

Not all emotions are universal. Some are highly specific to a culture or language – though they still capture human experiences that we can all identify with. Take *greng jai*, which in Thailand is the reluctance to accept somebody's help because it would cause them bother, *iktsurapok*, the Inuit name for the combination of fidgetiness and listlessness that sets in when waiting for visitors to arrive, or *awumbuk*, the Papua New Guinean sense of emptiness when they depart. Many northern European languages have a word to describe being cosy, warm and surrounded by friends. The Danes call it *hygge*, the Dutch *gezelligheid*.

deeper and know more. It can be provoked by an idea, a conversation, a creative project, a book or an interesting person. Like the big six, it has its own characteristic expression. As curiosity is aroused your head tilts to one side and the muscles in your forehead and around your eyes contract.

The flip side of curiosity – the sadness to its joy – is confusion. This, too, is common and easy to spot. The brow furrows, the eyes narrow and the lower lip might even get bitten. One study found it was the second-most recognisable everyday expression after joy. It is our brain's way of telling us that the way we are thinking about things is not working, that our mental model of the world is inadequate. Sometimes this will make us withdraw, but it can also motivate us to change our strategy.

Elevation

Another lesser emotion perhaps deserving of more attention in the modern world is 'elevation'. This is characterised by a tear in the eye, a tingling sensation on the back of the neck and a warm feeling in the chest. It can be inspired by a stirring speech, your national anthem or the sight of a loved one achieving something.

Elevation seems to be universal, having been documented in people from all over the world. And though it lacks an unmistakeable facial expression, it can be recognised by a subtle softening of the features and raised eyebrows, as if the person is sad. It is quite a rare emotion – people typically experience it less than once a week – but is highly motivational and significant. If you ask people to remember their most cherished experiences of their whole life, elevatory moments are likely to feature heavily.

On the negative side, there are shame, guilt and embarrassment, all of which serve to reinforce social bonds by signalling that we accept we have transgressed accepted norms. Pride also has a bad reputation – it is sometimes considered the worst of the deadly sins – but it motivates us to do things that gain respect.

If anyone ever accuses you of getting emotional, you can comfort yourself that emotions push us to act, and without them we'd never get anything done and society wouldn't function. So STICK THAT IN YOUR PIPE AND SMOKE IT.

Do our faces reveal our feelings?

In the late 1960s, a team of psychologists from the University of California in San Francisco travelled to Brazil, Japan, Borneo and New Guinea to show off a set of photographs. Some of the people they encountered had had no prior contact with the outside world and had never seen a photograph before. 'I was the first outsider they'd seen,' team leader Paul Ekman later recalled. But the photographs were still meaningful. They depicted the six basic emotions, explored in the last chapter: happiness, fear, anger, surprise, disgust/contempt and sadness.

Ekman's team found that everybody they tested, regardless of culture, recognised the same six. This classic research, published in *Science* in 1969, was hugely influential in psychology. It strongly supported Darwin's idea that expressions of emotions are common to all people, regardless of culture, because they have a common evolutionary origin.

Since then, dozens of studies have backed up the findings. In addition to the basic six (or seven, with contempt and disgust sometimes treated separately), the set of universal emotions has been extended to include pride, indicated by a tall posture and puffed-up chest, and shame, with a downturned head and bent posture. This all supports the view that emotional expressions are hard-wired into the human brain.

Innate reactions

Other support comes from studies of people who were born blind and have therefore never seen an emotional expression. A study of judo contestants in the 2004 Olympics and Paralympics, including athletes who were blind from birth or later became blind, found that all of them pulled the same faces when they won a bout. The expressions included what's known as Duchenne smiles – big, beaming smiles involving the eyes as well as the mouth – which are considered to be authentic expressions of happiness.

Social signals

Some emotional expressions, like fear and disgust, have obvious survival benefits for the person making them. Others are more mysterious. Psychologists are still working on potential utilitarian purposes of happy smiles, angry snarls and sad frowns.

It is possible that they don't have one, and merely evolved for signalling purposes. Humans are social animals who need to communicate, and facial expressions are a very powerful way of doing so. Being able to transmit and receive emotional states would have been advantageous to our ancestors. For example, displaying fear and reading

Ambiguous anger

Relying on facial expression alone can mean you badly misinterpret how somebody is feeling. One picture of tennis player Serena Williams is a good example. When you can only see her face she appears angry or in pain. But when you see her body too – on the front foot, fist clenched in front of her face – it is obvious that she is actually delighted. In fact, she has just beaten her sister Venus on her way to winning the 2008 US Open.

it on another person's face helps both of you to respond to danger.

In this scenario, emotional expressions started out as something evolutionary biologists call a 'cue' – they revealed information about an inner state or behaviour, but they didn't evolve for this signalling purpose, much as chewing is a reliable signal that somebody is eating. Then over time these expressions evolved into signals for expressly conveying information. This is a process called 'ritualisation'. Expressions became more exaggerated and distinctive, to make it easier to communicate non-verbally. This early stage of the evolution of expressions represents what's called an exaptation – a feature that initially evolved for one purpose but is co-opted for another.

This process may explain why it is hard to discern a function for some expressions: the original purpose has long been masked by ritualisation.

Dominance and submission

Another possibility is that some emotional expressions only ever served a signalling function. Pride and shame, which are particularly social emotions, are likely candidates. The expressions resemble the dominance and submission postures of other social primates, suggesting they are signals of status inherited from distant ancestors.

Ironically, the idea that facial expressions have diverged drastically from their original function has opened the door to the possibility that, contrary to Ekman's findings, they are not universal after all.

In other words, rather than being biologically based, they are culturally learned symbols – a form of 'body language' that we learn to communicate emotions. And, like spoken languages, the expressions share commonalities but also vary from culture to culture.

In support of this view, it turns out that people's recognition of an emotion depends heavily on context. In the real world, faces are rarely seen solely in isolation. Posture, voices, other faces and the wider context are also available for inspection, and they influence how expressions are perceived. For instance, a scowl – usually associated with anger – can be perceived as disgust if the person is holding a dirty object, or fear if it's paired with a description of danger. Similarly, viewing the same expression labelled alternately with the words 'anger', 'surprise' and 'fear' changes how people perceive it.

Cultural differences

Emotional expressions can also be culturally variant. Chinese people shown pictures of emotional expressions often categorise them differently from Europeans.

So if emotional expressions turn out to be less than universal, what does that mean? Ekman's research is often put forward as some of the best evidence we have for a universal human nature. So maybe deep down we're all different, and our differences are written all over our faces.

Crying and laughing
The Expression of the Emotions in Man and Animals is one of Charles Darwin's lesser-known books. Published in 1872, 13 years after *On the Origin Of Species*, it was daring for its time for its use of photographs. In it, Darwin argues that human expressions are connected to our mental states and evolved by natural selection. He commissioned a Swedish-born photographer called Oscar Rejlander to illustrate a variety of emotions. One photo really caught the public's imagination – a crying child mid-bellow. As a joke, Rejlander later sent Darwin this picture of himself with the photo, mimicking the child's face. To show how similar the expressions were for laughing and crying he added a second image of himself copying the child's laugh. Darwin's first edition sold a respectable 7,000 copies, but Rejlander really cleaned up, selling more than 100,000 prints from the book.

Credit: Reproduced by kind permission of the Syndics of Cambridge University Library (DAR 53.1)

Primordial fear in a modern world

December 26, 1973. All across the US, hordes of people brave the cold, dark winter to queue up outside movie theatres. Many wish they hadn't.

'I'm not going back in there,' said one woman after leaving halfway through. 'I just had to come out, I couldn't take any more,' said another. Some people vomited, others fainted. Cinemas started stocking smelling salts and barf bags.

The cause of their distress was *The Exorcist*, a movie about a girl possessed by an evil spirit. You might think that the scare stories would have driven people away, but all they did was add fuel to the fire. The film was a sensation and is still one of the highest-grossing films of all time.

Anyone who has seen *The Exorcist* – or *The Ring*, or *The Blair Witch Project*, or *A Nightmare on Elm Street*, or any one of thousands of similar movies – can identify with the audiences of 1973. Torn between watching and not watching, we peer between our fingers, waiting for the next stomach-churning jolt of fear, knowing it can only get worse. And when it's all over, we breathe a sigh of relief, laugh, and wonder why we did it.

It's a good question. Fear evolved for a reason – to make us run away, and stay away, from mortal threats. So why seek it out?

Appeal of the dread

Our fascination with horror is probably as old as our love of storytelling – which is to say, as old as human nature. *The Epic of Gilgamesh*, one of the oldest surviving works of literature, features a monster called Humbaba, created by the gods to be 'a terror to human beings'. Long before Humbaba was terrorising the ancients, people surely sat around campfires listening to stories of demons and monsters, casting nervous glances over their shoulders.

Today, darkness is still a time to close the curtains, turn off the lights and wilfully scare ourselves stupid. About a hundred horror, thriller and suspense movies are released every year, collectively grossing over $2 billion.

In academic circles, the appeal of horror was once explained in Freudian terms – as symbolic manifestations of repressed desires and anxieties.

The fearless few

A handful of people are unscareable. The most famous case is SM, who researchers spent six years trying, and failing, to scare. As well as showing her horror movies, they took her to an exotic pet store where she tried to pick up a snake and pet a tarantula. At Waverly Hills Sanatorium in Kentucky - a haunted house attraction billed as 'one of the scariest places on Earth' - SM just laughed. Her fearlessness is due to damage to a brain structure called the amygdala, which responds to threats and activates the fear program. Fearlessness may sound like a superpower, but is more like a curse. 'I wouldn't wish it upon anyone,' SM once said.

But that is giving way to an evolutionary account.

Our early ancestors lived in an environment replete with danger. Today, the savannahs of East Africa are patrolled by six species of large carnivorous mammal. Around 2.5 million years ago, there were many more, including sabretooth cats and giant hyenas.

Deep scars

There is little doubt that our ancestors were on the menu. In 1970, the 3.5-million-year-old skull of an Australopithecus child was discovered in Swartkrans cave in South Africa; it had a pair of circular puncture wounds matching the bite of a leopard. Fossil human bones have also been found with tooth and claw marks from lions, hyenas, crocodiles and even, in one case, an eagle. Modern hunter–gatherers continue to suffer high levels of predation. Among adult males of the Aché tribe, who live in the jungles of Paraguay, almost one in ten are killed by jaguars, usually at night.

This long evolutionary horror story has left deep scars in human cognition. Evolution has endowed us with a threat-detection system that is acutely sensitive to the mortal dangers our ancestors faced: snakes, spiders and, especially, large carnivores. Once triggered, the system initiates a sequence of physiological and emotional responses that prepare the body for escape or combat. The pupils dilate, the heart pounds, blood rushes to the muscles and blood-sugar levels rise. We experience this response as fear.

Hair trigger

The system is on a hair trigger, as the evolutionary cost of a false alarm is far lower than the cost of failing to respond. And so the merest hint of anything resembling a threat jolts it into action, even in modern city dwellers – a hosepipe in the grass, a centipede in the bath, the rustling of vegetation in the dark and, of course, fictional monsters.

Our imagination takes us beyond realistic monsters to include zombies, ghosts and other supernatural beings. But the underlying fear they evoke is the same: the mortal terror of predation. Whatever it is, it's watching us from invisible shadows, inscrutable, waiting to strike according to a schedule beyond our comprehension. We won't know what hit us until it is far too late.

Beneath the fear

What none of this explains is why people enjoy horror – or at least endure it. One of the best pointers comes from a woman known as SM, who is unable to feel fear due to brain damage. Shown clips from a series of horror movies, she showed no sign of fear, but she did find the clips exciting and entertaining.

That suggests that horror films are not merely scary: beneath the fear lurk other, more rewarding, emotions. That also makes sense from an evolutionary perspective. Although all humans have a general-purpose threat-detection system, our ancestors would have had to learn which specific things to fear in their environment. The legacy of this is that it is easier for city dwellers to acquire phobias of dogs, spiders and snakes than of genuine threats such as traffic and saturated fats. Those who found that learning process rewarding would be more likely to survive and pass on their genes. As a result, we also have an evolved tendency to perversely enjoy exercising our innate fear of predatory beasts, especially in an environment that we know is safe. Unless there really is something lurking under the sofa ...

Yeuuuugghhhh!

The word 'disgust' originally meant 'distaste', and you certainly wouldn't want to put many of the things that disgust humans into your mouth: faeces, slime, vomit, pus, rotting meat. But disgust is much more powerful than that. Disgusting things seem to have a force field around them, causing us to turn away, close our eyes and airways, and gag. 'Repulsion' would be a better word.

Not what it used to be

Disgust evolved to protect us from illness and death by keeping us away from harmful things. But when we became a super-social species its remit broadened. Nowadays it is also a very social emotion. It causes us to shun people who violate norms, or those we think, rightly or wrongly, are carriers of disease. As such, disgust is probably an essential characteristic for thriving on a cooperative, crowded planet.

It may also play a deeper and more insidious role in our everyday behaviour. Disgust has taken on a moral dimension too, offering a gut instinct reference for what is right and wrong. Numerous studies have shown that feeling temporarily disgusted increases how severely we judge people for things like shoplifting, political bribery or eating your dead pet dog.

Together, these findings raise all sorts of interesting, and troubling, questions about people's prejudices, and the ways in which they might be influenced or even deliberately manipulated. Humanity already has a track record of using disgust as a weapon against 'outsiders' such as immigrants, ethnic minorities and homosexuals. Nazi propaganda notoriously depicted Jewish people as filthy rats, and Hutu extremists in Rwanda stoked the genocide by referring to Tutsi as 'cockroaches'.

There is also experimental evidence that inducing disgust can cause people to shun minority groups, temporarily at least. In one experiment, a room was primed with fart spray before its occupants were invited to complete a questionnaire asking them to rate their feelings of warmth towards various social groups. One effect was stark: those in the smelly room, on average, felt less warmth towards homosexual men compared with participants in a non-smelly room. In another experiment, making people feel more vulnerable by showing pictures of diseases made them view outsiders less favourably.

More prone to disgust

Perhaps it's no surprise, then, to find that the more 'disgustable' you are, the more likely you are to be politically conservative. Similarly, the more conservative people are, the harsher their moral judgements become in the presence of disgust stimuli.

Studies like these suggest it might be possible to manipulate people's behaviour for the worse, for instance by strategically placing disgust triggers in or around a polling station. To an extent, many politicians have already come to the same conclusions. In April 2012, Republicans made hay of a story about Barack Obama eating dog meat when he lived in Indonesia as a child. Ahead of the primaries for the 2010 gubernatorial election in New York state, candidate Carl Paladino of the Tea Party sent out thousands of flyers impregnated with the smell of rotten garbage, with a message to 'get rid of the stink' alongside pictures of his rivals. Some political analysts believe his smelly flyers helped him thrash rivals to win the Republican nomination against the odds, though he lost the actual election heavily.

Given that disgust influences judgements of right and wrong, others have considered whether it might play a role in the courts. And it appears that it may well do. Experiments show that it clouds a juror's judgement more than anger, and that once people feel a sense of disgust, it is difficult for them to take into account mitigating factors. In court, disgusting crimes can attract harsher penalties: in some US states, the death penalty is sought for murders that have an 'outrageously or wantonly vile' element.

What sinks your boat?

It's amazing what you can learn to like. To the uninitiated, cheese is often disgusting, but Westerners have learned that it is tasty. Similarly, in Iceland rotten shark meat is a delicacy, and the liquor *chicha*, made from chewed and spat-out maize, is a popular drink in parts of South America. The influence of culture on disgust isn't limited to food. Kissing in public is seen as distasteful in India, whereas Brits are more repulsed by mistreatment of animals. Christian participants in one study even experienced a sense of disgust when reading a passage from Richard Dawkins's atheist manifesto *The God Delusion*. To a large extent, disgust is in the mind of the beholder.

Some say that instead of trying to overcome our sense of disgust, we should allow ourselves to be guided by it because disgust represents a deeper wisdom. For those still seeking to avoid its influence, it is worth noting that some people are more susceptible than others, and that the triggers vary according to culture. In general, women and young people tend to be more easily disgusted than men and older people. Women are also particularly sensitive to disgust in the early stages of pregnancy or just after ovulation – both times when their immune system is dampened and they are especially vulnerable to disease.

If they so choose, it is possible for anybody to become desensitised by continued exposure over time. Faeces are a potent disgust trigger, but it's amazing how easy it is to overcome that when you have to deal with your own offspring's bowel movements. Likewise, after spending months dissecting bodies, medical students become less sensitive to disgust relating to death and bodily deformity.

Drawing the sting

Other tricks include preventing people from curling their lips when they're grossed out. By simply asking them to hold a pencil between their lips, you can reduce their feeling of disgust when they see revolting images, and mollify their judgement of moral transgressions.

Happily, our lives are already a triumph over disgust. If we let it rule us completely, we'd never leave the house in the morning. The air we breathe comes from the lungs of other people and contains molecules of animal and human faeces. Everything we eat, drink and touch is coated in bacteria. It would be wise not to think about that too much. It really is quite disgusting.

The noxious side of nostalgia

For such a forward-looking species, we do like to wallow in the past. From endless movie remakes and costume dramas like *Downton Abbey* to politicians promising to turn the clock back to a simpler and better time, nostalgia sells.

Nostalgia is often seen as a harmless emotion – a sentimental longing for a past that is at best rose-tinted and at worst never existed. But that is to underestimate its power. Nostalgia can also be a powerful motivator. It is not just about longing for the past but is also a forward-looking force that shapes the future. Nostalgia can provoke political upheaval, xenophobia and bitter tribalism, yet it can also promote well-being, tolerance and a sense of meaningfulness in life. By better understanding its influence, we are now finding ways to harness its benefits and, just as importantly, anticipate its harms.

The word nostalgia – from the Greek *nostos*, to return home, and *algos*, pain – was coined by medical student Johannes Hofer in 1688, to describe a disorder observed in homesick Swiss mercenaries stationed in Italy and France. Hofer saw nostalgia as a disease whose symptoms included weeping, fainting, fever and heart palpitations. He advised treating it with laxatives or narcotics, bloodletting or – if nothing else worked – by sending the soldiers home.

While people's awareness of nostalgia of course pre-dated Hofer's label, his classification of it as a disease shaped how it was thought of for more than 150 years. One 1938 paper in the *British Journal of Psychiatry* referred to 'immigrant psychosis': a condition marked by a combination of homesickness, exhaustion and loneliness.

But by the second half of the twentieth century, the concept of nostalgia shifted. Rather than being pathological we now understand it as an emotion. At once a mixture of happiness and longing, its bittersweet nature is unique – but universal. Like many other emotions, it is universal across cultures, a part of human nature.

It is also surprisingly common. Some of us are more prone than others, but most people experience nostalgia at least once a week. That makes it all the more odd that we may have been getting the emotion wrong all this time.

Positive feelings

On a personal level, nostalgia tends to spring up when we are feeling low. But far from being the cause of sadness, it is an antidote. In general, it boosts well-being. Reflecting on nostalgic events you have experienced enhances positive feelings and self-esteem.

When people listen to songs that have particular meaning to them, the feeling of personal nostalgia increases perceptions of purpose in life. Pondering the point of your existence also causes a surge of nostalgia. One explanation for this is that nostalgia gives us a sense of stability and continuity. While so much in our lives can change – jobs, homes, relationships – nostalgia reminds us that we are the same person we always were.

Being nostalgic has been compared with applying for a bank loan. Looking back at our past is like checking our credit history: if the past was good, it suggests more of the same will follow.

In some ways, nostalgia is a by-product of how we remember. Memories are inaccurate: we routinely filter them to focus on the positive. Each time we recall something, we reactivate the memory, making it susceptible to alteration – a process called reconsolidation. So whenever we summon a memory, we might lose some nuances and add misinformation.

Nostalgic memory is all about the emotion, not what really happened. We might not remember the precise details of a momentous event such as the first day of college, but we broadly remember how we felt.

A sense of belonging

Nostalgia also serves a collective purpose. When a group shares a vision of the past – something known as 'collective nostalgia' – it promotes a sense of belonging and strengthens in-group bonds. This may have had survival benefits in early, tribal societies and can also be a positive force today. But often it comes at the cost of driving discrimination towards outsiders. When this type of nostalgia is tapped into on a national level, it increases nativism such as hostility toward immigrants. Our tendency to remember only the positive details of the past means we envision happier days when things were safe and secure, then seek to exclude people who were not a part of this carefree past – even if that past never existed.

But nostalgia alone is not to blame for surges in nationalism or anti-immigrant sentiment. Like any other emotion, it can be evoked for good or bad. Nostalgia can also promote social cohesion and charitable giving. When people think about a shared past experience with someone from a different background or stigmatised group, it improves attitudes toward those individuals. One study found that there were more donations when appeals on behalf of victims of an earthquake spoke of 'restoring the past' rather than 'building their future'.

Perhaps because politicians, advertisers, charities and social media have cottoned on to the power of nostalgia, there is a general sense that harking back to the past is more common than ever. And recent results suggest it is indeed on the rise, with more people noting that they, and people they know, are feeling nostalgic. It's enough to make you feel nostalgia for a simpler time when people were less misty-eyed about a past that never really existed …

How nostalgic are you?

There are different types of nostalgia, from longing for a simpler past to remembering your own life with a rosy hue. Psychologists measure personal nostalgia by having people rate how much they miss certain things from their past. The more nostalgic you are, the more likely you are to be emotionally intense, look for meaning in life, seek help from others in hard times and have strong relationship and coping skills. The less you are prone to nostalgia, the more likely you are to see yourself as very independent, feel less connection to others and use less healthy coping skills. Despite stereotypes about older people being most nostalgic, this emotion actually peaks in early adulthood. Men and women are also equally nostalgic.

Can money buy happiness?

Ask people if they think more money would make them happy and most answer emphatically 'Yes!' In reality high incomes don't increase happiness, but do create a life you think is better

Happiness increases with annual income, but only up to $75,000

Small garden

Fish pond

Ping-pong table

Small house

Kids go to state-funded school

Compact car

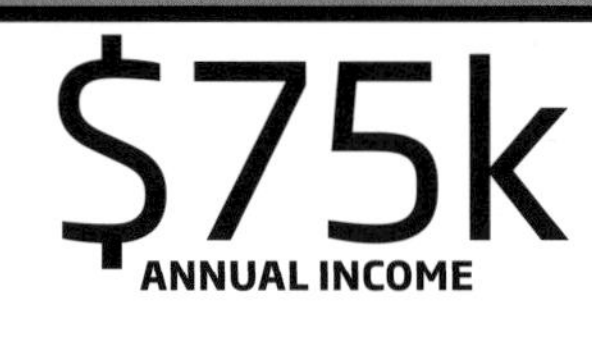

As income falls below $75,000, sadness, worry and stress increase. Poverty adds to life's other misfortunes, making them worse. So poor people suffer worse than the rich

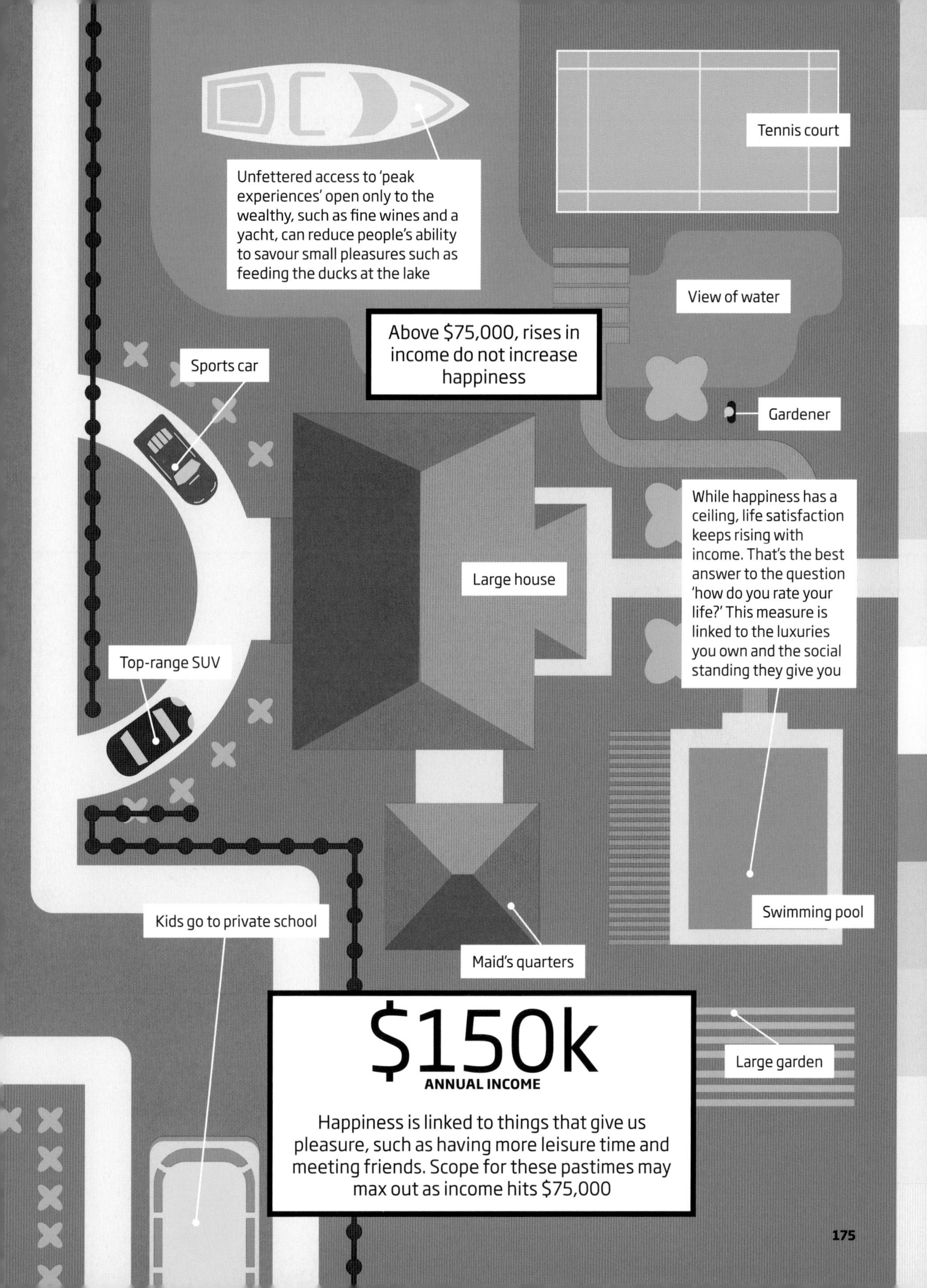
Unfettered access to 'peak experiences' open only to the wealthy, such as fine wines and a yacht, can reduce people's ability to savour small pleasures such as feeding the ducks at the lake
Tennis court
View of water
Above $75,000, rises in income do not increase happiness
Sports car
Gardener
While happiness has a ceiling, life satisfaction keeps rising with income. That's the best answer to the question 'how do you rate your life?' This measure is linked to the luxuries you own and the social standing they give you
Large house
Top-range SUV
Swimming pool
Kids go to private school
Maid's quarters
$150k
ANNUAL INCOME
Happiness is linked to things that give us pleasure, such as having more leisure time and meeting friends. Scope for these pastimes may max out as income hits $75,000
Large garden

Why do we cry?

When was the last time you cried? And what did you cry about? If you're a small child the answer is probably 'today' and 'because I hurt myself' or 'I got told off'. But for adults it is more complicated. Chances are the last time you cried had nothing to do with being sad or in pain.

Emotional crying is downright weird. Many animals produce tears to protect their eyes, but humans alone cry out of feeling. We cry not only for negative reasons, but also when we're happy, overwhelmed or enraptured. Why do we do it? More pointedly, when should you keep a stiff upper lip, and when might it help to turn on the waterworks?

People have long puzzled over crying. Aristotle purportedly viewed tears as an excretion like urine. Darwin concluded that, in addition to lubricating the eye, tears serve as a relief to suffering, although he didn't explain exactly how.

Let it out

The idea that crying is cathartic remains popular. But what does that even mean? For Freudians it suggests the release of pent-up emotions – the principle behind folk wisdom encouraging people to 'let it out'. Another interpretation is that crying rids the body of harmful chemicals, such as stress hormones, produced by emotional distress. It's a nice idea, but not very plausible. The average cry only produces around a millilitre of tears.

So why do many people say that crying makes them feel better? Perhaps it's simply because our mood improves by the time we've finished crying, even if it's just returning to normal from rock bottom. There is also some evidence that crying relaxes the body by activating the parasympathetic nervous system, or by pumping up levels of oxytocin, the 'cuddle hormone'.

However, the true function of crying is not

A lifetime of tears

How we cry, and what we cry over, changes as we age. For their first few weeks, babies don't even shed tears, because their tear glands are still developing. But as they grow, crying becomes less vocal and more tearful. Around adolescence, we begin to cry less over physical pain and more over emotional pain. Many people also start to exhibit 'moral crying' in reaction to acts of bravery, self-sacrifice and altruism. As we age, we increasingly shed tears over things that are positive. However, these 'tears of joy' may not reflect happiness at all; events such as weddings and holidays are often bittersweet because they remind us of the passage of time and mortality.

physiological, but social. Crying signals that we need help. Many animals clean their eyes and reduce irritation by secreting tears from the lacrimal glands, above the outer corner of each eye. Tears are also produced in response to damage and disease. As humans evolved and became more empathic, tears acquired a second role – as a cue for caregiving. And once crying started to elicit help from others, it became worth our while to shed tears over any hurt, physical or mental.

Still, why did the eyes become the channel for signalling distress, and not sweaty palms or pale lips? Well, the eyes are perhaps the best clue we have to what others are thinking, so we are predisposed to look at them. What's more, we can generally count on eyes to be visible.

Some of us are more inclined to using this signal than others. Aside from babies, neurotics and people who are highly empathic cry the most. Neurotics use tears manipulatively – as do narcissists, psychopaths and tantrum-throwing toddlers. Sociopaths are thought most likely to cry fake or 'crocodile' tears. And, although boys and girls cry frequently until puberty, in Western cultures women cry at least twice as often as men. There may be more to this than cultural conditioning: studies in animals suggest that the hormone testosterone suppresses tears.

Honest sign

Although crocodile tears are possible, one reason crying affects us so strongly is that it's hard to fake. It is considered an honest signal, which makes it really powerful. Seeing a picture of tears on a sad face for just 50 milliseconds boosts feelings of sympathy, supportiveness and friendship towards the individual pictured. Tears also help us overcome feelings of revulsion, which may make us more likely to help an injured person who is crying.

But our reactions to crying depend on a lot of factors – although not necessarily in the way gender stereotypes might dictate. For example, show people pictures of nurses and firefighters breaking down in tears while trying to help an injured person, and you might expect them to be more sympathetic to the nurses, given their stereotypically feminine role. In fact, it's the other way around. Crying is seen as more acceptable for a firefighter than a nurse. What's more, it makes no difference whether either is male or female.

So, doing a stereotypically masculine job may give you more of a licence to cry. But being female means you are more likely to be comforted when you cry. This could have an unexpected explanation rooted in how tears alter our looks. Men tend to think tears make faces look more masculine and also younger. So it has been suggested that tears may help protect women from inappropriate male advances and aggression – eliciting help instead.

Men may not be sending out the same signals when they cry. Indeed, in many cultures, they are under pressure to suppress their tears. Nevertheless, in certain domains, including competitive sports, crying is acceptable. And sometimes it can be downright desirable. For example, men who show intense yet controlled emotion in a sad situation are seen as more competent than those who show no emotion at all.

Blubbing isn't always the sign of weakness it's made out to be. But tears must be used wisely. How positively they are viewed depends on the context – which has to be perceived as important, and not your fault. How you cry also matters: welling up usually makes a better impression than openly weeping. Ironically, the powerful are more often admired for their tears than the weak.

Why boredom is positively stimulating

It is often said that only boring people get bored. If that is true, humans really are a tedious bunch. Around 90 per cent of people report being bored some of the time, and the average person says they experience about 6 hours of boredom every week.

Given boredom's dismal reputation, you might think that the 10 per cent of people who don't experience it are the lucky ones. But they might well be missing out on something. Boredom, it turns out, can be positively stimulating.

Four flavours of ennui

Everyone knows the feeling of boredom. The world around you dulls, concentration lapses, time starts to drag and all the things you could do seem equally unattractive and yawn-inducing.

But defining boredom so that it can be studied in the lab has proved difficult. For a start, it isn't simply about apathy, but can include a lot of other mental states, such as depression, frustration – the 'bored to tears' feeling – agitation and even a Zen-like indifference. There isn't even agreement over whether boredom is always a low-energy, flat emotion or whether it can also be energetic.

Perhaps it can be all of these things. Boredom has been classified into at least four different types: apathetic, searching, reactant and indifferent.

Of these, apathetic boredom is the closest to classical boredom. Not only are you listless, but unmotivated to do anything about it. Searching boredom is a more agitated, restless state, associated with active efforts to find something to do. But failure may lead to reactant boredom, an explosive combination of high arousal and negative emotion. Finally, there is indifferent boredom, which is relaxed, calm and not especially unpleasant.

Intriguingly, while most people experience all these kinds of boredom and might flit from one to another in a given situation, they tend to specialise in one.

Having defined boredom, the next question is, what is it for? This isn't as dumb a question as it might sound. Boredom is an emotion, and emotions generally serve a biological function.

The answer to this may lie with other animals. Like other emotions, boredom didn't just arise when humans came on the scene. Many other creatures, including mammals, birds and even some reptiles, seem to have a version of it, suggesting that there is some kind of survival advantage to feeling bored. The most plausible explanation is that it serves as an internal kick up the backside. Wild animals that have done nothing for a while often go out to look for things to do, which has clear survival value. Exploring territory, for example, can alert an animal to something useful or dangerous.

Get up and go

Human boredom may be more complex, but there are parallels. In his book, *Boredom: A lively history*, Peter Toohey at the University of Calgary, Canada, compares it to disgust – an emotion that motivates us to avoid certain situations. 'If disgust protects humans from infection, boredom may protect them from "infectious" social situations,' he suggests.

Where boredom stops being useful and starts becoming a problem is when the desire to explore is thwarted. That's why zoo animals in a plain wire cage can end up exhibiting strange behaviours such as pacing in a figure of eight or pulling out their own feathers. The human equivalent is probably being stuck in a traffic jam, unmoving train or airport departure lounge with no information about when something – anything – will break the tedium.

Boredom-busting exploration can take other forms too. People who are bored rigid by being made to spend 15 minutes copying out numbers from a telephone directory later perform better on a test of creativity. It may be that boredom allows – or even causes – the mind to wander in search of a creative solution.

Positive experience

Indifferent boredom, meanwhile, can be a positive experience. Someone who isn't engaged in anything satisfying but isn't fed up either can actually feel relaxed and calm. In the right circumstance – on a train ride home after a busy day at work or a family weekend, for example – this type of boredom can be extremely useful, helping us to unwind.

It is not all positive, however. People who get bored easily face poorer prospects in education, their career and life in general. They are also more likely to have problems with anger and aggression, and to partake in risky behaviours such as boozing and gambling.

One study even seemed to suggest that it's possible to be bored to death. Researchers looked at self-rated boredom levels in civil servants in 1985. When they followed them up in 2009, they found those who had been consistently bored were significantly more likely to have died early.

Of course, boredom itself cannot kill, it's the things we do to deal with it that may put us in danger. What can we do to alleviate it before it comes to that? The answer may be to tackle boredom head on – in other words, when you recognise that something is likely to be boring, get stuck in. People who approach life this way report less boredom than those who try to avoid it.

Born to be bored

The standard way to measure a person's propensity for boredom is the boredom proneness scale (BPS). This reveals that men get bored more than women, and extroverts more than introverts. People with narcissistic personality traits, anxious types and those who lack self-awareness are also susceptible. Competitive and thrill-seeking types are particularly prone, which has led some to suggest a link between boredom and a heightened desire for stimulation. On the other hand, creative people and those with a higher need for mental stimulation seem to be partly protected from boredom, perhaps because they do better at finding some interest or meaning in whatever task they have to do, however tedious.

Our techno-loaded, overstimulated lives might be part of the problem too. With so many distractions, we are neglecting our inbuilt boredom-buster, the ability to daydream. Our overconnected lifestyles might even be a new source of boredom as we flit from one app or device to the next seeking instant gratification.

Dorothy Parker once said 'The cure for boredom is curiosity. There is no cure for curiosity.' It's not boring people who get bored, but incurious ones.

A blow to the heart

'Sticks and stones may break my bones, but words will never hurt me.' Who wasn't offered those soothing words after being taunted at school or ridiculed in the playground? But as everyone knows all too well, they don't ring true. In fact, being hit by sticks and stones is often preferential to being dumped, ridiculed or betrayed. Emotional pain is exactly that: it *hurts*.

Sting of rejection

Our language has a rich seam of pain metaphors to describe emotional distress: we talk of being 'heartbroken', 'kicked in the teeth' or 'slapped in the face'. Such comparisons occur around the world: Germans talk about being emotionally 'wounded', while Tibetans describe rejection as a 'hit in the heart'. We now know that these turns of phrase capture something essential about the human condition. The sting of rejection fires up the same neural pathways as being hit with a stick.

The first hints that this was true came in the 1990s, when animal studies showed that morphine not only relieves pain after injury, but can also reduce the grief of rat pups separated from their mother.

Soon afterwards researchers put people in a brain scanner and deliberately subjected them to social rejection. Their instrument of torture was a computer game called Cyberball, in which three players pass a virtual ball among themselves. Volunteers were told that they were playing with two people who were in another room, but in fact they were controlled by the computer.

Although the game started out friendly, the computer soon stopped throwing the ball to the volunteer. A trifling insult for sure, but the volunteers' brains responsed as if they had been kicked in the teeth – and not metaphorically. The scanner revealed a surge in activity in the dorsal anterior cingulate cortex (dACC), a region known to be an important part of the brain's pain network which determines how distressing we find a physical injury. The more distressing you find an injury, the more the dACC lights up, a fact that also seemed to play out during the game: those who reported feeling worst after the rejection showed the greatest activity in this region.

Real pain

Other studies found that social rejection also provokes other parts of the pain network that are more directly linked to the feeling of pain, not just our response to it. Another slightly masochistic

Take care with teens

Being rejected or humiliated is painful at any time of life, but adolescents seem particularly susceptible to emotional sticks and stones. The brain's pain network is still developing at their age and, compared with the adult brain, it tends to show a more exaggerated response to slights and insults. On the positive side, social support during this period can carry lasting benefits. For instance, young adults who enjoyed tighter social networks in their late teens show more muted reactions to the sting of rejection.

experiment asked people who had recently been dumped to look at photos of their ex while recalling the gory details of their break-up. The same people were later subjected to a painful jolt of heat on the forearm. Their brains responded the same to both insults. As expected, the dACC lit up, but so did sensory centres associated with the actual sensation of pain. Heartbreak literally hurts.

Closest connection

Physical and emotional pain are so intertwined that they can feed off each other. When people feel excluded, they are more sensitive to burning hot probes, while submerging a hand in ice water for a minute leads people to report feeling ignored and isolated.

The converse is also true: the soothing of physical pain can alleviate the sting of an insult. People given a daily dose of painkillers over three weeks grew thicker skins, reporting less emotional distress in their daily lives, and fewer feelings of rejections when Cyberball partners excluded them (don't try that at home: prolonged use of painkillers can have harmful side-effects).

This all might explain why certain people find it harder to withstand the rough and tumble of social life than others. Extroverts have been shown to have a higher pain tolerance than introverts, and this is mirrored by their greater tolerance for social rejection. People who feel more pain when a hot electrode touches their arm are also more sensitive to hurt feelings in a game of Cyberball.

These differences may be partly genetic. People with a small mutation to the gene *OPRM1*, which codes for one of the body's opioid receptors, are more sensitive to physical pain. They are also more likely to slip into depressed feelings after rejection than are those without the mutation.

As with many traits, a child's early environment can determine their sensitivity. For instance, people with some forms of chronic pain are more likely to have had traumatic experiences, such as emotional abuse, during their early years. Perhaps it puts their pain network into overdrive, making them more sensitive to any discomfort.

When you consider our ancestors' dependence on social connections for survival, it makes sense for us to have evolved to feel rejection so keenly. Being kicked out of a tribe would have been tantamount to a death sentence. As a result, we needed a warning system to prevent us from causing further offence and teach us to toe the line in the future. The pre-existing pain network was ideally equipped to do the job and was co-opted by evolution.

Protective effect

Even in the modern world, emotional rejection can be unhealthy or even deadly. Large-scale studies have shown that people with good social connections are less likely to die than lonely people: the protective effect is on a par with abstaining from smoking or heavy drinking.

This may be because people who are lonely tend to have increased expression of inflammation genes, and chronic inflammation has been linked to a host of conditions, including heart disease, cancer and Alzheimer's. Lonely people are at a greater risk of all of these. In comparison, sticks and stones feel like child's play.

The three pillars of emotional wisdom

Some people are naturally good at navigating the world of emotions. We call them 'emotionally intelligent'. They are said to have a high EQ, in the way that a genius has a high IQ. But what if your EQ score is not so good? If you are an emotional dunce, or just a bit emotionally dim, is there any hope of improvement? The concept of emotional intelligence suggests not. Like regular intelligence, it's an innate ability – something we are born with.

But don't despair. As we discover more about emotions, this interpretation of EQ is looking increasingly rocky. It is becoming clear that all of us can improve our emotional competence by honing three key skills.

Primal forces

A popular view of emotions sees them as powerful, primal forces we struggle to understand both in ourselves and in others. The idea that emotions control us makes sense when you trace them back to their origins. Some evolved to help animals react quickly in life-or-death situations.

The fight-or-flight response is a classic example. Before you are conscious of feeling fearful, your body and mind are already primed to act – your heart is racing, your vision focused, and you experience a hot rush of blood to the head and perhaps an urge to lash out.

Emotions generate these sorts of physiological changes in all animals. But for us they are more than just subconscious calls to action. We also have many social emotions such as jealousy, sympathy and guilt. They are what make our emotional lives so complicated, and some people are clearly better at coping with this complexity than others. But this is not an innate ability. It's a skill that can be learned. Emotional competence is a language, in which we can become more fluent. Just as learning a language entails recognising words, understanding how to use them, and controlling a conversation, so mastering the language of emotions requires three key skills – perception, understanding and regulation of emotions.

Spot the emotion

Perception is the bedrock on which the two other skills rest. Perceiving emotions is not as straightforward as it might seem. For a start, facial expressions of emotions are dynamic. Computer-generated faces that randomly combine actions, such as lip curls and raised eyebrows, reveal that each emotion has an associated sequence of facial movements, which unfolds a bit like the letters of a word. Strung together in specific patterns, these create 'sentences' that communicate a more complex social message. Then there are other visual signals in our gestures and movements. And all these can interact with aural cues such as tone of voice and other sounds.

The Geneva Emotion Recognition Test (GERT) is a way of assessing how good people are at perceiving emotions in everyday life. It involves a series of short videos of actors expressing an emotion by uttering meaningless syllables. Scores can range between 0 and 1, with an average of around 0.6. People who get higher scores on the test are both better negotiators and perceived as being nicer and more cooperative than people with lower scores. But GERT also offers a way to improve your skills. When volunteers first learned the cues in face, voice and body associated with different emotion, and then practised using the video clips and getting feedback, the average score was 0.75.

Musical training may help too. We know that adult musicians are better than non-musicians at judging the emotion in someone's tone of voice.

And music training also modulates brain responses associated with emotions and with our ability to interpret others' minds.

Interpret the signs

Recognising emotions is not enough, though. You also have to understand how they are used – and that's the second skill. Not everyone smiles when they're happy, or scowls when they're angry. What's more, brain scans show tremendous variability in brain activity both between people and in the same individual, in response to different types of threat. So, to correctly grasp emotions, you need knowledge and flexibility.

Understanding emotions in this way is learnable rather than innate. A programme called RULER, used in some 10,000 US schools, teaches children and young adults to interpret physiological changes in their bodies and label the linked emotions. Other researchers are investigating whether having a broad and accurate vocabulary for your own emotions can make you more aware of other people's.

Once you can recognise and make sense of emotional signals, then you need the final skill – the ability to regulate your feelings. Again, this isn't something we are born with. What's more, as we develop, some of us learn ineffective strategies for doing it, such as avoiding emotionally charged situations or trying to shut down our emotions completely.

Temper your response

There are ways to improve your regulation skills. 'Reappraisal' involves trying to put yourself in someone else's shoes so as to be more objective, and change your emotional response accordingly. Hairdressers, waiters and taxi drivers taught this strategy found that it resulted in more tips. Another promising approach is mindfulness – observing the coming and going of your emotions without action or judgement. This strategy has been linked with increased job satisfaction and reduced emotional exhaustion. Seeing emotions as thoughts and sensations, provides a sense of perspective and the 'hot' aspect of emotions dissolves.

Other animals may be slaves to their emotions. Human emotional life is more complex and cerebral. Learn to master it and you will reap the rewards that emotional competence brings.

The language of emotions

The facial expressions of pure, unfettered emotions are the same for us all, even people who have been blind from birth. Yet we also regulate our emotions to conform to cultural norms. If you live in a culture where anger is viewed as disturbing and selfish, for example, you will not be rewarded for expressing it, and over time you may even cease to feel it as frequently or intensely. This will also affect the way you interpret anger in others. Immigrants gradually adapt their emotions to the norms of their new home. It's as if we all speak the same language but adopt the local dialect.

9
Life Stages

Waaaaaahhhhhhh!

As anyone who has ever had one knows, human children take a long time to pop out and are then small, helpless and extremely demanding. But they are nothing compared with some of our close relatives

ADULT SIZE

NEWBORN SIZE

Birthweight 2.5 tonnes

This is unusually variable

Red kangaroo

Gestation: 33 days

1/100,000 adult weight

The roo is born very undeveloped. It crawls the short distance into its mother's pouch, attaches to a teat and continues to develop. After about 200 days it starts leaving the pouch for increasing amounts of time. 50 days later it leaves permanently

Echidna

22 days

1/10,000 adult weight

The female lays a single soft-shelled egg directly into her pouch. 10 days later the fetus-like baby (called a puggle) emerges. It remains in the pouch for another 50 days or so. The mother then deposits it into a nursery burrow, returning every 5 days or so to suckle. After a further 140 days the mother digs the young out of the burrow, stops suckling, and leaves it to its own devices

Giant panda

95-160 days

1/800 adult weight

Cubs are born tiny, blind, hairless and helpless. They are the smallest mammalian newborns apart from the marsupials. They open their eyes after 6-8 weeks, are weaned at 6 months but remain dependent on their mothers for up to 2 years

Blue whale

10-12 months

1/70 adult weight

Calves can swim as soon as they are born. At 7 metres long they are larger than some fully-grown whales of other species. Weaned at 6 months but remain with their mother for up to 3 years

1 gram

0.3 grams

100 grams

Largest animal of all time

The largest mammal baby in proportion to its mother's bodyweight

African elephant

20–21 months

1/45 adult weight

Calves are born well-developed and can walk almost immediately. They depend on their mother's milk for 2 years and often continue to suckle for 5 or more, until they are too tall to reach her teats easily

Human

9 months

1/20 adult weight

Babies are born almost totally helpless and depend on parental care for 14–16 years

Giraffe

14 months

1/10 adult weight

Newborn calves fall to the ground from a great height but are protected by an unusually tough birth membrane. Calves can walk almost as soon as they are born. They suckle for a year

90 kg

3/5 kg

100 kg

Why we all suffer childhood amnesia

We've all been there. Shame no one can remember a thing about it.

As babies, we are phenomenal learners. In our first couple of years we pick up many complex, lifelong skills, like the ability to walk, talk and recognise faces. Yet most of us remember nothing from these early years. It's as if someone has torn the first few pages from our autobiography.

This puzzling phenomenon, known as 'childhood amnesia', is universal. Most people recall nothing at all before their third birthday, though some people claim to remember events before age two, while others draw a blank until they were six or even eight. And those early memories are hazy. Not until middle childhood do we recall anything with clarity.

Freud's thinking

What causes childhood amnesia? Sigmund Freud believed that we repress early memories because they are full of sexual and aggressive impulses too shameful for us to face. That idea eventually fell by the wayside, to be replaced by the view that young children just can't form explicit memories of events. Then, the first studies of children themselves – rather than investigations of adults' childhood recollections – revealed that children as young as two or three can recall autobiographical events, but that these memories fade away. The question therefore became: what causes early memories to disappear?

The anatomy of the brain probably plays a part. Two major structures are involved in the creation and storage of autobiographical memories: the prefrontal cortex and the hippocampus. The hippocampus is thought to be where details of an experience are cemented into long-term memory. And one small area of this region, called the dentate gyrus, does not fully mature until age four or five. This area acts as a kind of bridge that allows signals from the surrounding structures to reach the rest of the hippocampus, so until the dentate gyrus is up to speed, early experiences may never get locked into long-term storage.

Yet children can still remember some events before this region is fully developed, so it can't be

Whose memory is it?

You have heard that adorable anecdote from your childhood a million times. You can see the scene clearly in your head. But is your recollection real, or have you concocted a false memory around an oft-told family tale?

We can't wholly trust any of our memories – they always contain missing information and misremembered details. And experiments show that we are particularly susceptible to creating false memories relating to events during the period of childhood amnesia. That could have important bearings for court cases that rely on early memories, such as those investigating allegations of childhood abuse.

the be-all and end-all of childhood amnesia. What else is involved?

At around 18 to 24 months of age, just before autobiographical memory begins to surface, toddlers reach a key milestone. They start to recognise themselves in a mirror. This is a sure sign that they now have a sense of self – the understanding that the entity 'me' is different from 'you'. That ability helps us to organise our memories, making them easier to recall. Yet that can't be the whole story either, because memories continue to be sparse well beyond the point at which our sense of self appears. Growing evidence reveals the extra ingredient to be language.

If children aged between two and four are given a novel toy to play with, then asked six months later to describe it, they use only words that were part of their vocabulary when they first played with the toy. Even though their vocabulary has grown by leaps and bounds in the meantime, it's as though the memory is locked into their language at the time of the event. What's more, adults' earliest memories associated with words such as 'ball' or 'Christmas' tend to date from a few months after the average age at which each cue word is acquired. In other words, you must have a word in your vocabulary before you can set down memories of that concept.

Language matters

If a sense of self provides a structure around which to organise memories, the development of language provides a further kind of memory scaffold. Language allows a child to construct a narrative upon which to anchor the details of an event. The importance of narrative is apparent in a study that recorded how mothers spoke to their children when they were between two and four years old. Ten years later, children whose mothers had used more elaborate language had earlier memories than those whose mother's conversations were more repetitive.

It would appear that the way we talk to our kids when they are young shapes what they remember years down the road. This could explain puzzling cross-cultural differences in the age of earliest memories. In one study, for example, researchers found the average age of first memories in people of European descent hovered around 3.5 years, compared with 4.8 years for east Asians and 2.7 years for Maori people in New Zealand.

Elaborate storytelling

Compared with east Asian parents, European and North American parents tend to discuss the past more often with more elaborate storytelling. As a result, their children have more early memories. The Maori storytelling culture is even richer, with detailed oral histories and a strong focus on the past, leading to even earlier memories.

Talking about the past doesn't just help children develop narrative skills, it also fosters development of a sense of self. So it looks as if language and self-perception go hand in hand, and both are necessary for autobiographical memory to flourish.

One big question remains: is it possible to reclaim memories from that period of our early childhood that is hidden from us? There's no doubt that very young children remember a lot in the short term. These recollections are fragile and may never become locked into permanent storage. However, it's possible they are retained but not accessible using traditional prompts such as words and images. If so, with more imaginative cues such as smells, flavours and music, we may one day be able to excavate our buried childhood memories.

What's the point of childhood?

As any parent will tell you, human childhood is a long haul. Gorillas are all grown up at six and chimps are ready to reproduce by the age of eight. At that age, humans are barely halfway through their childhood. Even elephants and blue whales are sexually mature by their early teens. We spend more time as children than any other animal on Earth, and continue to be dependent on our parents well into early adulthood.

Our extended childhood has intrigued researchers for decades. Some believe that it is simply a by-product of our long lifespan and requires no special explanation. But many anthropologists are convinced there must be some evolutionary advantage. Humans live complicated lives: we have culture, language and technical skills. We can build cities and civilisations, and survive on our wits and intellect. Perhaps our childhood is so different because we have so much more to learn than other animals?

One way to answer the question is to look at when childhood evolved. Our 3-million-year-old ancestors the australopithecines seem to have had no real childhood – like nonhuman apes, they moved swiftly from infancy to adolescence. It was not until about 1.5 millions years ago, with the appearance of our genus *Homo*, that childhood began to lengthen. By the time *Homo sapiens* appeared, about 100,000 years ago, childhood was the extended affair it is today.

The appearance of a long childhood coincides with a rapid increase in brain size. Could the two be connected? It can't simply be about giving our brains longer to grow, because they stop growing way before adulthood. Perhaps, then, it provides the time required for configuring the hardware to help us cope in the adult world.

One of the most obvious features of childhood is play. Play consumes most of a young child's waking hours, and even once formal schooling starts they still play whenever they get the chance.

Skills workshop for kids

Playing may seem like pure fun, but for children it is a serious business. Play is how children acquire many of the skills that will be indispensable to them as adults, such as how to interact socially, control their emotions, solve problems and innovate. These skills are so important and so complex that we need an extended period with few responsibilities to create the space and freedom to acquire them.

Play is not unique to humans, of course. It is an important part of the healthy cognitive development of many animals, and being deprived of play can be detrimental. For example, rats raised without access to playmates have severe cognitive and social deficits.

But humans are unique in that we engage in fantasy play: 'what if?' and make-believe games. Much of children's play involves pretending that one thing or person is another, for example that a cardboard box is a car, or a friend is a mother. This ability is thought to be how we develop our unique symbolic abilities including self-awareness, language and theory of mind.

In support of this, our extinct close ancestors the Neanderthals had much shorter childhoods than us. Despite almost certainly possessing language, the symbolic culture they left behind was much less rich than ours.

Play is also crucial for creativity. Research on social carnivores such as wolves, coyotes and dogs suggests that it lets individuals try out new things in an environment where they can make mistakes without much penalty.

Creativity requires the creation of many novel ideas, and the ability to try them out in different

Children have their uses

Strange as it seems, our extended childhood may have evolved for the benefit of adults. Other mammals continue suckling their young until they can fend for themselves, which limits the number of offspring. Small humans, in contrast, can rely on other family members for food and help, which frees their mothers to become pregnant again. Modern parents burdened by the cost of children may baulk at the idea of having ever more mouths to feed, but the idea of kids as 'dependent' is a modern concept. Across the world, children typically look after their younger siblings and do household chores while consuming few resources. For parents in these cultures, childhood may be too short rather than too long.

combinations and spot useful consequences. These are all embodied in play.

Dolphins provide a vivid example. One of their most creative approaches to feeding is to drive fish to the surface with bubbles produced deep underwater. This is a variation on games they play with bubbles. It's difficult to imagine how dolphins would have discovered this if not through the accidental consequences of play.

As for dolphins, so for humans. There are many examples of people attributing major achievements to play, including Nobel laureates Richard Feynman and Alexander Fleming. 'I play with microbes,' Fleming once said, by way of a job description.

Absorbing knowledge

Another way in which childhood is an extended apprenticeship for adulthood is through formal education. Not only do we have to master social skills, we also have to acquire knowledge. Again, our long childhood creates time and space to do so.

Our ancient ancestors were hunters and gatherers, occupations that need specialist skills and knowledge. Hunters must learn how to make weapons, locate and track animals, kill and butcher them. Gathering plants also takes skill. Foragers search out hard-to-obtain foods such as tubers, nuts and honey and often collect material that must be processed before it can be eaten.

Modern education serves a similar function. Why else do kids spend years in school, studying and acquiring skills, if not to help them function effectively as adults?

Unfortunately, the balance between play and education seems to have shifted too far. Research from countries such as Sweden and Finland, where the early school years are dedicated to informal play-based learning, overwhelmingly shows that delaying education until age seven works better than imposing it at age four. Children in these countries have better academic achievement and well-being, despite 'missing out' on 2 or 3 years of formal schooling. Teaching five-year-olds to read may even damage their literacy in the long run. As the old saying goes, all work and no play makes Jack a dull boy.

Can we blame teenage torment on the brain?

It's not easy being human. Just when you get the hang of being a child, you suddenly start to change into an adult. The transition from one to the other can be bewildering, exhilarating, infuriating and emotionally draining. And that's just for the parents.

No other species has teenagers. Even our closest relatives, the great apes, move smoothly from juvenile to adult life. So why do humans spend half a decade or more in the teenage twilight zone, with all the accompanying gormlessness, moodiness, recklessness, acne and disturbed sleep patterns?

Traditionally, the teenage years have been seen as a sort of reproductive apprenticeship, but it may have at least as much to do with shaping our brains.

The first adolescents

Evolution offers two big reasons for thinking this might be true. The first relates to when adolescence evolved. Evidence from growth in the bones and teeth of fossilised hominins indicates that it emerged sometime between 800,000 and 300,000 years ago – only a very short time before our ancestors' brains underwent the last big expansion to reach today's size. The second indication comes from neurobiology and brain imaging, which show that there is a wholesale reorganisation of the brain during the teenage years.

Psychologists used to explain the particularly unpleasant characteristics of adolescence as products of raging sex hormones, since children reach near-adult cerebral volumes before puberty. Recently, though, imaging studies have revealed a gamut of structural changes in the brains of teens and early twenty-somethings that go a long way to explaining the tumultuous teenage years.

Brain scans that were carried out every two years on a group of 400 children, from adolescence into adulthood, show how the teenage brain upgrades itself to become quicker by a process of pruning and insulating. The changes take place in the outer layer of the brain, known as the cortex. As adolescence progresses, the cortex gets thinner, probably because unwanted or unused connections between neurons are pruned back. On average, teens lose about 1 per cent of their grey matter every year until their early twenties.

The cerebral pruning trims neural connections that were overproduced in the childhood growth spurt, starting with the more basic sensory and motor areas. These mature first, followed by regions involved in language and spatial

Teen owls

The late nights and long lie-ins favoured by your average teen are not just down to partying hard and bone idleness. As puberty begins, we turn into night owls: our biologically driven bedtimes and waking times get later. The trend continues until 19.5 years in women and 21 in men, then reverses. At 55 we wake at about the time we woke prior to puberty, which on average is two hours earlier than adolescents. This means that for a teenager, a 7 a.m. alarm call is the equivalent of a 5 a.m. start for a person in their 50s. No wonder they're so grumpy.

orientation and lastly those involved in higher processing and executive functions.

Among the last to mature is the dorsolateral prefrontal cortex at the very front of the frontal lobe. This area is involved in control of impulses, judgement and decision-making, which might explain some of the less-than-stellar decisions made by your average teen. This area also acts to control and process emotional information sent from the amygdala – the fight or flight centre of gut reactions – which may account for the mercurial tempers of adolescents.

As grey matter is lost, the brain gains white matter, a process called myelination. White matter, or myelin, is the fatty tissue that surrounds neurons, helping to conduct electrical impulses faster and stabilise the neural connections that survive pruning. Some see this as a trade-off. By pruning connections, we lose some flexibility in the brain, but the result is a more efficient brain – as ready as it will ever be to take on the world.

Risky behaviours

There are a few notable pitfalls to the maturing process. The lack of impulse control combined with less brain activity in areas responsible for risk assessment may lead to behaviours such as drug and alcohol abuse, smoking and unprotected sex. Substance abuse is particularly concerning, as brain imaging studies suggest that the motivation and reward circuitry in teen brains makes them almost hardwired for addiction. Throw in a lack of impulse control, poor judgement and a woeful underappreciation of long-term consequences and you have a hooked teen.

Compared with the adult brain, our teenage brain also tends to show a more exaggerated response to slights and insults. Social support can help with this, though. Young adults who had enjoyed tighter social networks in their late teens show subdued reactions to rejection compared with those who had felt lonelier through their teens, perhaps because memories of past acceptance subconsciously soothe their feelings.

Adolescent brains also make decisions and handle other people's feelings very differently to adults. Adults use their prefrontal cortex to answer decision-making questions, but teens use an area called the superior temporal sulcus. This region processes very basic behavioural actions, unlike the prefrontal cortex, which is involved in complex functions like understanding how our decisions affect others. So if you were under the impression that teenagers aren't very good at understanding the consequences of their actions, there may be a simple explanation.

As if dealing with a major brain reorganisation weren't enough, teenagers may also have a handicap when it comes to learning. Our ability to learn new languages or our way around a new location drops when we hit puberty. Mouse studies suggest this may be due to a temporary increase in a chemical receptor that blocks activity in a part of the brain involved in learning. Their number soars when mice hit puberty, then falls back again during adulthood. But there's light at the end of the foggy-brain tunnel: injecting pubertal mice with the stress steroid THP appears to compensate for their learning deficits. Remember though – this is in mice. Don't think of injecting your teen with stress hormones just yet.

Flying rite of passage

What a way to end your childhood! This is how boys become men on Pentecost Island in the Pacific archipelago of Vanuatu. Once a boy has been circumcised at seven or eight, he is free to take part in the ritual of 'land diving' – said to be the inspiration for bungee jumping. Land diving entails leaping from a 30-metre wooden tower with a carefully measured tree vine tied around each ankle. When a boy leaps for the first time, his mother holds a favourite object from his childhood. Once he lands safely, she throws the object away to signal the end of his childhood. A good dive is not only an expression of masculinity, it's also believed to ensure a plentiful yam harvest.

Credit: Raul Touzon/National Geographic/Getty

Does having kids make you happier?

Kids – who'd have them? Most people, actually. By the time we hit our mid-forties, a large majority of us have reproduced at least once.

Having kids is supposed to be a joyous, life-affirming experience. The reality is often quite different, and across the Western world record numbers of people are remaining childless. In the UK, one in five women have no children by the age of forty-four. In the US, the picture is similar for both genders, and the number of childless women has almost doubled since the 1970s. While many people may want kids but can't have them, some are simply rejecting what was once considered an inevitable and essential part of the human experience – procreation.

What price children?

In some ways, that's not so surprising. Having children can have a significant impact on finances, careers and the planet. The average middle-class US family has spent more than $245,340 on each child by the time they're eighteen. In the UK, the cost of raising a child has swelled 63 per cent in a decade, with childcare alone eating up 27 per cent of the average salary. Children, though small, also come with a large environmental footprint. The United Nations projects that if current population and consumption trends continue, humanity will need the equivalent of two Earths to support itself by 2030.

More surprising, though, is the impact of children on the health and well-being of their parents. Contrary to what we might think, study after study has shown that having children does not make people happier, and may even reduce happiness. Becoming a parent is associated with depression and sleep deprivation, leaves couples less happy with their sex lives and hastens marital decline. A study of more than 14,000 couples found that mothers reported a sharp rise in stress after the birth of a child – three times that of the father – and that it increased year-on-year. US mothers rank childcare 16th out of 19 everyday tasks in terms of positive feeling, just ahead of commuting to and from work, and work itself. Another study found that the average hit to happiness exacted by the arrival of an infant is greater than a divorce, unemployment or the death of a spouse.

On this basis, it might seem utter folly for couples to take the parenthood plunge. But can the situation really be that gloomy? There is some evidence that happiness and parenthood can coexist. One study found that having children made men (but not women) happier. However, this effect may have been largely due to the fact that most of the parents were married, which is known to increase happiness. Another study seemed to confirm the idea that people with kids are happier than people who don't have kids. But when it accounted for other differences between the two groups in things like wealth and religion, the boost to happiness disappeared.

Costs and benefits

Such niggles show just how complex parenthood and happiness are to study. Kids can't be handed out at random to see what effect they have on people. One way around this problem is with before-and-after studies of the same people. This approach has shown that parents' happiness increases a year or so before the birth of the first child, and then returns to pre-birth levels by the time the baby is about one.

So the true picture is clearly more nuanced than a blanket, 'Kids make you (un)happy'. There are many factors at play. One is money. There's evidence that parenthood tends to boost people's satisfaction with their lives apart from their financial circumstances – but for most people, the money woes associated with children are so great that any additional happiness they feel is swallowed up.

An upside to ageing

A parent's age matters too. For people younger than thirty, children are associated, on average, with a decrease in happiness. From thirty to thirty-nine, the average effect on happiness is neutral, and at age forty and above, it's positive. For them, it's the more, the merrier, to a point – three seems to be the optimal number.

Where you live can also make a difference. The happiest parents over forty live in former socialist states such as Russia and Poland, where care of the elderly falls mostly to the family, so having children is a boon in later life. Parents in the twenty to twenty-nine age group tend to sustain a large hit to their happiness by having children, but the generous welfare systems in countries such as Sweden, Japan and France soften the blow.

The happiness gap

Comparing happiness levels between parents and non-parents within a country, and then between countries, can serve as a sort of global barometer. In the US, where help for parents is extremely limited, the happiness gap between people with and without children is wider than in the majority of twenty-two other countries studied. Even if children don't bring happiness, you might still expect them to bring a sense of purpose. But that's not the case in the US and probably other countries where social and financial support for parents is lacking.

Parenthood, then, would seem to be a lottery. If you're lucky enough to be married, well-off, or a resident of a country with generous social provision, you have a better chance of enjoying it. For the rest, it may not be the experience they had hoped it would be. It's a testament to human optimism – or perhaps our bad decision-making – that the vast majority of people still would like to have children anyway.

Childbirth changes men

The physical effects of parenthood are not confined to mothers. Fathers' testosterone levels are lower than those of their childless peers, and lowest of all in dads who spend 3 or more hours a day caring for their child. This change is thought to allow men to switch from mating mode – where testosterone-fuelled competitiveness and musculature are an advantage – to parenting mode, where caring, attentive behaviours are important to reproductive success.

Parenthood can change a man's brain too. Scans taken between two weeks and four months after their child's birth reveal increases in grey matter in areas associated with parenting behaviours such as responding to a baby's cries.

What's the point of middle-aged people?

Life, it is sometimes said, begins at forty. But it often doesn't feel like it.

To many people, the realisation that youth is behind them and middle age has arrived is a depressing one. We are used to dismissing our fifth and sixth decades as a negative chapter in our lives – perhaps even cause for a crisis. But it shouldn't be seen that way. Recent scientific findings have shown just how important middle age is for every one of us, and how crucial it has been to the success of our species. Middle age is not just about starting our descent down the other side of the hill towards the inevitable. It is an ancient, pivotal episode in the human lifespan, pre-programmed into us by natural selection – an exceptional characteristic of an exceptional species.

Two healthy decades

Compared with other animals, humans have a very unusual pattern to their lives. We take a very long time to grow up, we are long-lived, and most of us stop reproducing halfway through our lifespan. A few other species have some elements of this life plan, but only humans have distorted the course of their lives in such a dramatic way. Most of that distortion is caused by the evolution of middle age, which adds two healthy decades after the babies stop appearing – two decades which most other animals simply do not get.

An important clue that middle age isn't just the start of a downward spiral is that it does not bear the hallmarks of general, passive decline. Most body systems deteriorate very little during this stage of life. Those that do, deteriorate in ways which are very distinctive, are rarely seen in other species, and are often abrupt.

For example, our ability to focus on nearby objects declines in a predictable way – long-sightedness is rare at thirty-five but universal at fifty. Skin elasticity also decreases reliably and often surprisingly abruptly in early middle age. Patterns of fat deposition change in predictable, stereotyped ways. Other systems – notably cognition – barely change.

Pre-programmed process

Each of these changes can be explained in evolutionary terms. In general, it only makes sense to invest in the repair and maintenance of body systems that deliver an immediate fitness benefit – that is, help to propagate your genes. As people get older they no longer need spectacular visual acuity or mate-attracting unblemished skin. Yet they do need their brains, and that is why we still invest heavily in them during middle age.

As for fat – that wonderfully efficient energy store which saved the lives of many of our hard-pressed ancestors – its role changes when we are no longer gearing up to produce offspring, especially in women. As the years pass, less fat is stored in depots ready to meet the demands of reproduction – the breasts, hips and thighs – or under the skin where it gives a smooth, youthful appearance. Once our baby-making days are over, fat is stored in larger quantities and also stored more centrally, where it is easiest to carry about. That way, if times get tough we can use it for our own survival, thus freeing up food for our younger relatives.

These changes strongly suggest that middle age is a controlled and pre-programmed process – a process not of decline but of development.

When we think of human development, we usually think of the growth of a fetus or the maturation of a child into an adult. Yet development, and the genetic processes which direct it, does not stop when we reach late teens or early twenties.

David Bainbridge is clinical veterinary anatomist at the University of Cambridge, UK, and author of *Middle Age: A Natural History*

It continues well into adulthood. The tightly choreographed transition into middle age is a later, but equally important, stage of human development when we are each recast into yet another novel form.

That form is one of the most remarkable of all. It is an evolutionary novelty unique to humans – a resilient, healthy, energy-efficient and productive phase of life which has laid the foundations for our species' success. Indeed, the multiple roles of middle-aged people in human societies are so complex and intertwined, it could be argued that they are the most impressive living things yet produced by natural selection.

Evolutionary objection

The claim that middle age has evolved faces one obvious objection. For any trait to evolve, natural selection has to act on it generation after generation. Yet we often think of prehistoric life as nasty, brutish and short. Surely too few of our ancestors lived beyond forty to allow features of modern-day middle age to have been selected for?

This is a misconception. Although average life expectancy may sometimes have been very low, this does not mean that *Homo sapiens* rarely reached the age of forty during the past 100,000 years. Average life expectancy at birth can be a misleading measure; if infant mortality is high then the average is skewed dramatically downwards, even if people who survive to adulthood subsequently have a good chance of living a long, healthy life.

That doesn't mean life wasn't nasty, brutish and short at times – especially after humans underwent the transition to agriculture between 12,000 and 8,000 years ago. But apart from this time, in which the longevity of adults may actually have decreased for a while, the evidence from skeletal remains suggests that our ancestors frequently lived well into middle age and beyond, and many modern hunter–gatherers live well beyond forty.

The probable existence of lots of prehistoric middle-aged people means that natural selection had plenty to work on. Those with beneficial traits would have been more successful at nurturing their

The middle-aged brain

Middle age may be the time when we start to experience senior moments, but loss of brain power isn't a major worry. A study of more than 7,000 civil servants suggested that there is a small cognitive diminution during middle age. However, whether this matters is debatable. Middle-aged people also tend to be better at developing long-term plans, selecting relevant material from a mass of information, planning their time and coordinating the efforts of others – a constellation of skills that we might call wisdom.

Indeed, functional brain-imaging studies suggest that they sometimes use different brain regions than young people when performing the same tasks, raising the possibility that the nature of thought itself matures as we get older.

children to reproductive age and helping provide for their grandchildren, and hence would have passed on those traits to their descendants. As a result, modern middle age is the result of millennia of natural selection.

Late developers

But why did it evolve as it did? The answer is inextricably bound up with the exceptional nature of humans. Our survival is entirely dependent on skilled gathering of rare, valuable resources. We cooperate, plan and innovate in order to extract what we need from our environment, be that roots to eat, hides to wear or rare metals to coat smartphone touchscreens. We lead an energy-intensive, communication-driven, information-rich way of life, and it was the evolution of middle age that supported this.

For example, hunter–gatherer societies often have complex techniques for finding and processing food that take a long time to learn. There is evidence that many hunter–gatherers take decades to learn their craft and resource-acquiring abilities may not peak until they are over forty.

Gathering sufficient calories is crucial for the success of a human community, especially since young humans take so long to grow up. Indeed, during the early years of life they devour calories without contributing many to the group. A human child requires resources to be provided by more than two young parents. For example, a study of two groups of South American hunter–gatherers suggested that each couple requires the help of an additional 1.3 non-reproducing adults to provide for their children. Thus, middle-aged people may be seen as an essential human innovation, an elite caste of skilled, experienced 'super-providers' on which the rest of us depend.

Crisis, what crisis?

One of the most enduring concepts related to middle age is the midlife crisis. It is usually associated with men in their early to mid-forties, and is often said to include a psychological crisis of self-worth, a tendency to seek the romantic attentions of inappropriately young women, and a reversion to interests such as sports cars. However, it may not be a real phenomenon at all.

None of the three elements stands up to scrutiny. Men in their forties are no more likely to claim they are experiencing a crisis than men or women in their thirties or fifties, and many researchers now believe that the concept of the midlife crisis should be consigned to the dustbin.

The other key role of middle age is the propagation of information. All animals inherit a great deal of information in their genes; some also learn more as they grow up. Humans have taken this second form of information transfer to a whole new level. We are born knowing and being able to do almost nothing. Each of us depends on a continuous infusion of skills, knowledge and customs – collectively known as culture – if we are to survive. And the main route by which culture is transferred is by middle-aged people showing and telling their children what to do, as well as the young adults with whom they hunt and gather.

These two roles of middle-aged humans – as super-providers and conveyers of culture – continue today. In offices, on construction sites and on sports pitches around the world, we see middle-aged people advising and guiding younger adults and sometimes even ordering them about. Middle-aged people can do more, earn more and, in short, they run the world.

Subtle changes

This has left its mark on the human brain. As might be expected of people propagating complex skills, middle-aged people exhibit no dramatic cognitive deterioration. Changes do occur in our thinking abilities, but they are subtle. For example, response speeds slow down over the course of adulthood. However, speed isn't everything, and it is still debated whether other abilities deteriorate at all.

A central and related feature of middle age is the many healthy years we enjoy after we have stopped reproducing. Female humans are especially unusual animals because they become infertile halfway through their lives, but male humans often also effectively 'self-sterilise' by remaining with their post-menopausal partners. Almost no other species does this.

The possible benefits of the menopause are not immediately obvious. Natural selection favours individuals who rear the most offspring. Yet there are other, rare examples of reproductive cessation in the animal kingdom which may provide some clues. Orcas also undergo menopause, and it is striking how much their lives mirror ours. They are long-lived, slowly developing, intelligent and vocally communicative. They invent and apply a complex array of techniques for communal food acquisition and they are extremely widespread.

Thus, humans can be seen as members of an elite club of species in which adulthood has become so long and complicated that it can no longer all be given over to breeding. Just like long-sightedness and inelastic skin, the menopause now appears to be a coordinated, controlled process. Recent research suggests that it is not a meandering, stumbling deterioration but a neatly executed event that is a key part of the developmental programme of middle age. It liberates women and their partners from the unremitting demands of producing children, and gives them time to do what middle-aged people do best – live long and pamper.

Challenging time

Middle age is fascinating because it links our species' history to individual people's experiences. Each person is destined to pass through a phase of life for which they frequently do not feel prepared. The sudden end of fertility challenges their self-image, their appearance alters before their eyes, and even the ways their brain works change. The midlife crisis, middle-aged parenthood, the empty nest syndrome and new, unexpected urges all beckon, but at last science has started to tackle the once-inexplicable forces behind these phenomena.

Few people look forward to middle age. Some fear it, some joke about it. Yet recent advances in palaeoanthropology, neuroscience and reproductive biology are revealing the truth about this long-neglected phase of human life. Without the evolution of middle age, human life as we know it could never have existed.

Teaching old dogs new tricks

People in their sixth or seventh decade often decide it is time to learn something new. Some try yoga, others take up cookery or enrol on a degree course. But very few have expectations of getting seriously good at their new skill. As hair turns grey and waistlines expand, our mental cogs supposedly start to seize up, making it difficult to pick up new skills. You can't teach an old dog new tricks, right? Or maybe you can.

A decade ago, few neuroscientists would have agreed that adults can match the learning talents of children. But we needn't be so pessimistic. The mature brain remains surprisingly supple, and the idea that youngsters are inherently better at learning is on its way out.

Stuck in a rut?

The notion that the mind gets stuck in a rut as it ages is culturally entrenched. The 'old dog' aphorism is recorded in an eighteenth-century book of proverbs. When scientists began to investigate the malleability of the adult brain in the 1960s, their findings appeared to agree. Most insights came indirectly from studies of perception, which suggested that visual abilities could only develop during a critical window in infancy. For example, a young animal blindfolded for a few weeks after birth will never develop normal vision. The same is true for people born with cataracts or a lazy eye – intervene too late, and it cannot be fixed.

The assumption was that other kinds of learning were similarly constrained. Consider learning a second language. Anecdotally, this gets harder as you age. The young children of immigrants pick up their adopted tongue with ease while their older siblings struggle and their parents get nowhere.

However, the evidence that this is really the case turns out to be weak. US census records, for example, detail the linguistic skills of immigrants. If there really was a critical period for learning a second language, it should show up in the data. But researchers found no discernible difference between people who moved to the US in early childhood and those who did so in adolescence. Even people who moved as adults were only marginally less fluent.

Language acquisition is a special case, however, and there is still the suspicion that children might have the edge in skills involving perceptual or motor learning: things like singing or playing a new sport. Learning involving these abilities differs from the acquisition of factual knowledge, because it needs us to rewire the eyes, ears and muscles. Yet the available evidence suggests that children are not inherently superior at this either.

Nor are adults necessarily slower at picking up the intricate movements that are crucial for music or sport. In one challenging test of hand–eye coordination, nearly 1,000 volunteers of all age groups learned to juggle over a series of six training sessions. People aged sixty to eighty began with some hesitation, but they soon caught up with the thirty-year-olds and by the end of the trials all the adults were juggling more confidently than the five- to ten-year-olds.

More time and attention

The reason that children appear to be better learners may have more to do with extrinsic factors. Children spend almost all of their waking hours learning to do things, both in formal tuition and in experimental play. They also receive feedback, praise and admonishment that are rarely dished out to adult learners for fear of patronising or offending them. If adults had that kind of time and attention, they would surely learn just as effectively. Indeed, experiments suggest that adult learners given the same kind of feedback as schoolchildren are actually better learners.

A glut of free time and a carefree existence are out of reach for most of us, but there are techniques that can be integrated into an adult's schedule. For example, children are continually quizzed on what they know, and for good reason: testing doubles long-term recall, outperforming all other memory tactics. Yet most adults attempting to learn new skills rely more on self-testing.

On tasks that involve motor skills or perceptual learning, adults can hamper their progress with perfectionism. Whereas children throw themselves in at the deep end, adults often agonise over the mechanics of the movements, trying to conceptualise exactly what is required before attempting to do it. Or they worry about looking foolish. This overthinking is one of the biggest barriers to learning.

Don't linger too long

Excessive conscientiousness also gets in the way. Adults are better than children at devising and sticking to practice regimes, but these can backfire. Left to their own devices, most adults segment their sessions into blocks. When learning salsa dancing, for instance, they may work on a specific move until they feel they have mastered it, then move on to another. The approach may bring rapid improvements at first, but a host of studies have found that it is less effective overall.

Instead, you'd do better to take a carousel approach, rotating quickly through the different skills to be practised without lingering too long on each one. Although the reason is unclear, it seems that jumping between skills makes your mind work a little harder when applying what you've learned, helping you to retain the knowledge in the long term – a finding that has helped people improve in activities ranging from tennis and kayaking to pistol shooting. Whatever new trick you want to learn, being an old dog shouldn't stop you.

Stay fit to stay sharp

The key to a spry mind in later life can be as simple as a walk in the park or some other form of gentle exercise. Poor physical fitness can be as damaging to our brains as it is to our sex appeal, reducing the long-distance connections between neurons and shrinking the hippocampus, which is involved in learning and memory.

Thankfully, the decline can be reversed by exercise. Not only does it restore the hippocampus and long-range connections, it also improves attention, which should aid learning of any new skill.

The upside of old age

There's not a lot you can do to avoid getting older, except die young. That may not sound like much of a choice, but old age isn't really that bad. Yes, there's the inevitable physical and mental decline, and the knowledge that the end of life is a lot nearer than the beginning. But there are many upsides too – not least that there will be a lot of people your own age to hang around with.

At the start of the twentieth century, average lifespan in the West was in the mid-forties. That has risen to about eighty today. Much of the initial rise came from dramatically reduced infant mortality, but from the 1970s onward, further increases in life expectancy have been driven by older people dying later. This is mainly thanks to better healthcare, such as widespread use of medicines to lower blood pressure and cholesterol levels. To put it in perspective, of all the people in human history who ever reached the age of sixty-five, half are alive now. In 2009, the number of pensioners in the UK exceeded the number of minors for the first time in history.

Oldest of the old

That's remarkable enough, but what is even more so is that the real population explosion has been among the oldest of the old – centenarians. In fact, this is the fastest-growing demographic in much of the developed world. In the UK, their numbers have increased by a factor of 60 since the early twentieth century. And their ranks are set to swell even further: by 2030 there will be about a million worldwide. Perhaps you will be among them.

On the face of it that sounds like a society you might not want to be part of. Old age brings an increased risk of chronic disease, disability and dementia. Growing numbers of very elderly people can only mean more human suffering and a greater burden on society, right? This is the dark cloud outside the silver lining of increasing longevity. Yet researchers who study the oldest old present a less bleak vision.

Physical elite

It is becoming clear that people who break through the ninety-plus barrier represent a physical elite, markedly different from the merely elderly who typically die younger than them. Far from gaining a longer burden of disability, their extra years are often healthy ones – not only do they have a long lifespan but a long 'healthspan' too. Centenarians typically live quite independently until they are ninety-five or so. Supercentenarians – people aged one-hundred-and-ten or over – are even better examples of ageing gracefully, often living without nursing care until they are well into their 11th decade.

For some reason they have a remarkable ability to delay or entirely escape a host of diseases that kill off most of their peers – around 60 per cent of them either avoided the chronic diseases of old age until after eighty, or dodged them altogether. These 'escapers' hit their century with no sign of disorders such as heart disease, cancer, diabetes, hypertension or stroke.

The trend is particularly apparent for cancer. The odds of developing it increase sharply as people age, but they fall from the age of eight-four, and plummet from ninety onwards. Only 4 per cent of centenarians die of cancer, compared with 40 per cent of people who die in their fifties and sixties.

All told, centenarians spend fewer days ill and bedridden than younger elderly groups, and when the end comes it comes quickly.

This is good news from both personal and societal perspectives, for it means that exceptional

Time to retire

One of the benefits of getting older is the prospect of quitting work and going on holiday for the rest of your life. For that we have to thank the German chancellor Otto von Bismarck. In the 1880s he needed a starting age for paying war pensions. He chose the age of sixty-five because that was typically when ex-soldiers died. But today in developed countries, people can reasonably expect to live for another 15 years. That is a hell of a long holiday. But thanks to increased longevity, retirement in your mid- to late sixties may soon be a luxury we can no longer afford.

longevity does not necessarily lead to exceptional levels of disability, or long drawn out years of ill health before death. It also provides us with clues about how more of us might achieve a long and healthy old age.

Secrets of a long 'healthspan'

So what are the secrets of a long and healthy life? Some people simply get lucky in the genetic lottery. Take a close relative of a centenarian and you can put money on their chances of living a long life. Among Americans born in 1900, brothers of centenarians were 17 times as likely to reach a century as their peers –and sisters, eight times.

But genetics isn't even half of the story. Gerontologists also point to four key factors: diet, exercise, 'psycho-spiritual' well-being and social connectedness. Around 70 per cent of longevity is due to these non-genetic factors, all of which are possible to cultivate.

Alzheimer's is rare

Another way to avoid the ravages of old age is to keep mentally active. Neurodegenerative diseases are common among centenarians, with around 80 per cent suffering from some form. But Alzheimer's disease, the most common form of dementia, is relatively rare.

Intriguingly, autopsies of the oldest old often reveal extensive brain lesions that are usually associated with Alzheimer's disease, even though the person had shown no outward sign of dementia. This resistance to Alzheimer's is often seen in people who lived intellectually stimulating lives, which suggests it, too, can be cultivated by lifestyle choices (see 'Protecting your brain from the ravages of time', p 206).

Ultimately, of course, the end will always come. The oldest confirmed human on record was Jeanne Calment, a French supercentenarian who died in 1997 aged 122 and 164 days. The fact that her record has not been broken since suggests that she was pretty close to a fundamental upper limit on human longevity. Gerontologists generally agree that – radical life-extension technology notwithstanding – going much beyond 120 is impossible. But look on the bright side. Your days may be numbered, but the number can be a lot bigger – and better – than you thought.

Protecting your brain from the ravages of time

One of the most painful aspects of watching Alzheimer's disease in action is its agonisingly slow pace. Bit by bit, it demolishes memories, personality and independence to leave a shell of the person you once knew. The process can take years.

But it doesn't always happen that way. Scientists have long been intrigued by a group of people who show little if any mental deterioration until just before they die. Autopsies often reveal their brains to be riddled with the plaques and tangles characteristic of advanced Alzheimer's.

Protection from dementia

People who die in this abrupt fashion tend to be better educated, more intelligent, lead more intellectual lives, and have high-status occupations. They appear to be protected from not only Alzheimer's but also the mental decline that comes with age and other damage from stroke, head injuries, HIV, alcohol abuse and Parkinson's disease.

About 25 years ago scientists put this cushioning effect down to what they called 'cognitive reserve'. The more you have, they argued, the more brain damage you could sustain without showing signs of mental decline.

Initially the idea was contentious. Some argued that it was stating the obvious that people who start off smarter have further to fall. Others noted that intelligence and socio-economic status tend to go hand in hand with better general health. Today, however, we have plenty of evidence that cognitive reserve is a real thing. What's more, we have a good idea of the biology that underlies it. For anyone who hopes to live a long, healthy and mentally sharp life, the implications are huge.

The first hint that some people have added mental resilience emerged in 1992 when researchers studied the blood flow in the brains of Alzheimer's patients. Outwardly, all showed symptoms of equal severity, yet those who had received more education had more severe brain pathology. Some kind of 'padding' seemed to be shielding these patients from the mental deterioration that would have been expected from the state of their brains.

Since then, evidence has piled up in support of cognitive reserve. Better-educated people, for example, suffer a smaller amount of cognitive decline for a given level of damage to their white matter, which is linked to mental decline in old age. They are also less likely to experience a marked decline in IQ after a head injury.

So cognitive reserve is real. But biologically speaking, what is it? The obvious answer is that it's all down to brain size: if you have more neurons, you should be better able to cope with losing some

Decline and fall

It may seem strange that people with a large cognitive reserve should go downhill so quickly when diagnosed with Alzheimer's disease. What's thought to be going on is that by the time these people show any symptoms of decline they are already at a relatively late stage of the disease. Their cognitive reserve masks the outward symptoms for longer than would normally be expected. But once the reserve is used up, their fall is dramatic.

of them. Sure enough brain size does correlate with cognitive reserve, but not strongly enough to be the only contributor.

Then in the mid-2000s, researchers used magnetic resonance imaging to watch the brains of people with Alzheimer's as they performed a cognitive task. They discovered that highly educated people are better at recruiting alternative networks of neurons to compensate for deterioration in cortical areas that handle complex behaviour and thought. Put simply, their brains have added flexibility.

There may well be another contributor to cognitive reserve. Scan the brains of people as they perform increasingly difficult memory tests and you find that the higher a person's IQ, the less effort their brains have to make to complete the tests. Their brains are more efficient at processing information, which may make them more resilient in the face of age-related degeneration or disease.

Building a buffer

Clearly, cognitive reserve is something we all want. So how can we get it and can we add to it?

One of the most important factors is, of course, intelligence, which is partly decided by our genes. But intelligence is not immutable. For example, a good predictor of people's intelligence in later life is their IQ when they're around ten years old. But studies show that plenty of elderly people have significantly higher IQs than would have been expected from those childhood scores.

One reason for this uplift is almost certainly education. This is borne out by research which shows that one of the strongest predictors of cognitive ability at age fifty-three is a person's educational attainment at twenty-six. In other words, education through the teens and early twenties can make a big impact on our mental skills, even much later on. Education is also likely to train people to recruit those alternative neural networks that are so important to cognitive reserve.

An active mind

If education boosts cognitive reserve, so does a track record of performing mentally demanding tasks. Intellectually stimulating jobs have been shown to provide a degree of protection against the risks of dementia and Alzheimer's.

For people who missed out on education and a mentally challenging job, there are other options. Evidence suggests that staying mentally active helps to cushion people who have started to suffer age-related decline. Crosswords and puzzles can stretch the mind, and reading has a positive effect on Alzheimer's symptoms, reducing their severity. Then there's the mental dividend paid by keeping physically active: exercise not only helps to stave off decline but can boost memory (see 'Faster body, faster mind', p 230).

Which of these activities works best at building cognitive reserve is still to be tested. But the best advice is 'do something': cognitive function is modifiable throughout life and it's never too late to take control.

That's important because beyond sixty-five, a person's risk of dementia roughly doubles every five years; already around one-third of people over the age of eighty-five have it. At a time when the number of elderly people is rising fast, building cognitive reserve promises huge benefits in reducing the burden on hard-pressed health systems and easing personal suffering.

10
Sex and Gender

How do men and women really differ?

As any fan of pop psychology will tell you, men are from Mars and women are from Venus. That isn't meant to be taken literally, of course, but it captures our everyday experience of sex and gender. We are one species, divided (roughly) into two tribes that sometimes seem to come from different planets. But are men and women really that different?

The basic biology of sex differences is well established, and largely boils down to sex chromosomes. Around half of fertilised eggs have two X chromosomes and half an X and a Y (the casting vote comes from the sperm that wins the fertilisation race; the female gamete always contributes an X). To a first approximation, XX zygotes are destined to become girls and XY boys.

To begin with, all start out more or less female. Unless interrupted by hormones, the default developmental pathway is towards a female anatomy. But sometime between 6 and 12 weeks of pregnancy a gene on the Y chromosome switches on and triggers the production of testosterone. This causes the genitals to develop into a penis and testicles rather than vagina, clitoris and ovaries.

Pro-female genes

That is not the whole story. There are also 'pro-female' genes that switch on in XX fetuses and promote the development of female anatomy. For example, a gene called *R-spondin 1* promotes the development of the ovaries. And once puberty kicks in, the sex hormones testosterone and oestrogen drive the development of secondary sexual characteristics such as breasts in women and deep voices in men. The genitals also develop further.

So far so uncontroversial. The vast majority of people are either XX or XY and have an anatomy to match. But fascinating as our anatomical differences are, they are not the be-all and end-all of what divides the sexes. Males and females also differ in their behaviour and psychology.

Battle of the sexes

The evolutionary roots of sex differences can often be found in sexual selection – the proliferation of traits, such as the peacock's tail, that are considered most attractive by members of the opposite sex.

Take competitiveness. Male-male competition is a near-universal feature of traditional human societies, and winners reap rewards in terms of social status and desirability to women. The same competition exists in industrialised societies, though is often a proxy contest for wealth.

Sexual selection has also shaped women, but for different traits. These differences inevitably lead to antagonism. The simple fact is that a woman must nurture a fetus for nine months whereas a man's input to procreation can last a few minutes. Such differences can explain diverging priorities and attitudes in key areas such as parental investment and number of sexual partners.

This, of course, is contentious stuff, suffused by sexual politics and arguments about nature and nurture. One of the most controversial debates is over differences between male and female brains (see 'Blue brain, pink brain', p 214). Nonetheless, scientists have developed a good picture of which behaviours show sex differences, and to what extent.

Gender divides

Let's start with something pretty uncontroversial. Men are, on average, taller than women. The mean height of British men is about 175 centimetres; the equivalent figure for women is 162 centimetres.

This average difference can be captured by a number called the standard deviation unit, which takes into account not just average height but also its variability. For height, the standard deviation is about two.

That doesn't mean that men are always taller than women. It allows for the fact that many women are taller than men – enough so that height alone is not a reliable predictor of an individual's sex. An individual who is 180 centimetres tall is more likely to be a man than a woman, but you can't say for sure.

With that benchmark established, how do other traits measure up? Only two characteristics show a greater divide: gender identity and sexual orientation. For the vast majority of people, gender identity matches their biological sex: most anatomical males think of themselves as male, and most anatomical females think of themselves as female. Likewise, most people prefer sexual partners of the opposite sex. On the standard deviation scale, gender identity differs by around twelve units and sexual orientation by six.

That does not mean that all anatomical males identify as male or prefer to have female sexual partners, or vice versa. It just means that gender identity and sexual orientation are even stronger predictors of an individual's sex than height.

Next on the scale is play behaviour, which differs by about the same amount as height (it is slightly stronger in males than in females). In practice this means that boys are on average more likely than girls to engage in rough-and-tumble play or to choose a truck over a doll, but there are enough exceptions to this rule that it is not possible to predict a child's sex from play preferences alone.

Whether this difference is down to innate biological differences or gender stereotyping is highly debatable. It is probably a bit of both: male vervet monkeys prefer cars even though they have never been gender-stereotyped, and girls who have a hormonal disorder that means they produce abundant testosterone prefer them, too. After that, most of the differences are only around half the size of those on height, which means they are not very large at all. Nevertheless, there are real and measurable differences.

The traits where men score higher (on average) than women are aggression, assertiveness and the ability to mentally rotate objects. Women, meanwhile, outdo men on empathy, fine motor skills, perceptual speed and verbal fluency.

Further down the list comes mathematical attainment, which shows far less male bias than we are often led to believe. Right at the bottom of the table are numerous traits commonly assumed to be skewed by sex but which in practice show no discernible difference between men and women. These include computational skills, overall verbal ability and leadership potential.

So, yes, men and women are different. But only on average, and biology is not destiny. Forget Mars and Venus – we're all citizens of the same planet.

Kingdom of women
Mosuo women celebrate new year near Lugu Lake in the Tibetan Himalayas. This ethnic group has a population of about 50,000 and is unusual because it is largely ruled by women. Children trace their lineage back through the female side of the family. Men are in charge of livestock, fishing and politics, but women run just about everything else; they head up households, manage the money, assign jobs to family members and pass property to their daughters.

Credit: Patrick Aventurier/ Gamma-Rapho via Getty

Blue brain, pink brain

Imagine a young couple driving around a city they don't know, trying to navigate their way to a wedding. The one in the passenger seat fails to read the map properly and they end up lost. The driver refuses to stop to ask for directions, insisting that they can find their way. The passenger gets in a flap and starts to cry. The driver's temper flares, and they end up arguing.

In your imagined scenario, who was who? Chances are you pictured a woman in the passenger seat and a man behind the wheel.

If so, you can't be blamed. Our culture is awash with gender stereotypes that are hard to ignore. Perhaps the most pervasive is that fundamental differences in behaviour are hard-wired into our heads. Brains come in hues of either pink or blue, as one researcher puts it.

Diverging brains

The divergence of the female and male brains supposedly begins in the womb. We know that male fetuses are anatomically masculinised by sex hormones. The assumption is that testosterone also moulds their developing brains, carving out lifelong differences in the architecture of various neural circuits. Divergence continues throughout childhood, with both nature and nurture playing a role. By the time we reach adulthood there are numerous structural differences between the brains of men and women, at all levels of organisation, from gross anatomy to circuitry, signalling chemicals and synapses.

All told, there are something like 30 regions of the brain that differ between men and women. One of the most famous is a region involved in spatial reasoning tasks such as mentally rotating three-dimensional figures (no prizes for guessing which sex has a larger one). Men also have larger

Wired for trouble

There are clear differences in the types of mental illness and learning difficulties that males and females experience. Boys are much more vulnerable to developmental disorders such as Asperger's syndrome, dyslexia and ADHD. Major depression is twice as common in women, while men are more susceptible to alcohol dependence and antisocial personality disorder.

These sex differences may be the result of different vulnerabilities due to the distinct ways in which those brains are wired. We know, for example, that the amygdalae, important for processing fear and aggression, are bigger in men, while the hippocampi, critical for memory, are bigger in women.

amygdalae, a pair of almond-shaped structures deep in the brain that process emotions such as fear and aggression. Women, on the other hand, appear to have larger areas of the brain associated with language and memory.

It is tempting to conclude that these account for those stereotypical differences between male and female behaviour. Men are better at map-reading, DIY and violence; women possess superior social skills and are excellent multitaskers.

But not so fast. A common criticism of the 'blue brain, pink brain' idea is that there is only a small average difference between the sexes, with more variability within each sex than between them. In other words, the results tell us about averages across the population, not individual differences.

Indeed, when neuroscientists look at individual brain scans, they find that very few people have a typical male or female brain. Most of us have a mix of male and female brain features. This means that, even though average sex differences in brain structure do exist, an individual brain is likely to be just that: individual, with a mix of features.

In any case, the behavioural stereotypes are just that: stereotypes. In terms of cognitive skills and personality characteristics, the two sexes are much more similar than different. Just knowing whether someone is male or female is a very poor predictor of almost any kind of behaviour.

In fact, the idea of blue and pink brains may be entirely the wrong way round. The physical disparities between male and female brains are not what creates differences in behaviour or ability, but are there to iron them out.

Nobody doubts that males and females must be genetically and hormonally different, in order to create two sexes with different sets of genitals and reproductive behaviours. Male brains are awash with testosterone, for example, while female ones experience monthly cycles of the hormones oestrogen and progesterone. These may be necessary to help males and females develop their different reproductive strategies, but may produce unwanted consequences outside of behaviours related to reproduction. Sex differences in the brain may have evolved to compensate.

In support of this, brain-scanning studies have found differences in the workings of male and female brains that are not accompanied by differences in their performance.

Compensating for sex differences

In one study men and women were asked to name everyday objects as they flashed up at a rapid pace. There was no gender difference in the scores, but the male brains were working much harder, showing much more activation in brain regions thought to be responsible for visual recognition. It is possible that men's brains have to compensate for their relatively weaker language abilities.

More evidence of compensatory circuits at work involves the amygdalae. Even when the brain is at rest, amygdala activity is different in men and women. The difference could be a compensatory mechanism to make up for differences in testosterone levels.

If a neuroscientist was given someone's brain without their body or extral information, they would still probably be able to guess if it had belonged to a man or a woman. Men's brains are larger, for example, and are likely to have a larger number of male features overall. But it is impossible to predict what mix of brain features a person is likely to have based on their sex alone. The idea of pink and blue brains is not a total myth, but most of us have a bit of both.

Going dating? Don't forget the peacock's tail

Playing the dating game can be the most exhilarating and rewarding adventure. Or the most terrifying and humiliating. Magazines and books are filled with ways to attract a mate, most of them wild speculation.

Fortunately, science has plenty to say on the topic. Biologists have been fascinated by the way animals compete for a partner ever since Charles Darwin proposed his theory of sexual selection. He used the peacock's tail to illustrate his idea. A fan of fabulous feathers doesn't help a bird to stay alive, yet the fact that he can thrive with such an extravagant appendage suggests he is fitter and has better genes than other males. Peahens respond accordingly.

Humans are no different. The clothes you wear on a date, your chat-up lines and flirting style are equivalents of the peacock's tail: evolved for reproductive success. Adjusting them to take account of evolution's lessons might give you the edge in enticing a mate.

Choose your colour

Let's start with the colours you wear on a date. Female animals often use red to signal fertility, and there's plenty of human folklore and cultural references linking red with sex, passion and fertility. Men clearly recognise this link: they sit closer to women who wear red, rather than other colours, and ask them more intimate questions.

Red also seems to work for men. Some male primates flush red after a surge of testosterone. The male hormone dampens immunity so a red individual is letting females know he's fit enough to deal with this shortfall. Women who see men in a range of coloured shirts rate those in red as more desirable and of higher status.

If red is not your colour, there are other ways to increase your allure. For a man, one option is to look like you're off the market. In general, women are more likely to pursue a man if he has a partner than if he is single.

Women also rate a man more highly if they see a woman smile admiringly at him, or if he is surrounded by women. This effect does not work when the tables are turned, however. Men consistently rate a woman in a group of other men as less appealing.

What lies beneath

Why these preferences exist is up for debate. It may be that a woman who entices a man away from his partner feels she's a better, more attractive person. Alternatively, women may favour an attached man because he's already been chosen as a good mate, while the qualities of a single man are unknown. The male instinct to avoid women who are popular with other men may be born of the desire not to be cuckolded and left to raise another man's child.

For men and women alike, being noticed is one thing, but at some point prospective partners have to talk. And here, chat-up lines can be thought of as the peacock's tail. The opening exchange gives players the chance to flaunt their qualities.

Women tend to prefer men with higher IQs, for example. Smarter men may not necessarily be better breadwinners, but they do tend to be healthier and have higher sperm quality. So a chat-up line that shows off intelligence should be more successful than a lewd remark or compliment.

Despite this, men still persist with lines like: 'I may not be Fred Flintstone but I can still make your bed rock!' Why do they do this? One suggestion is that such lines force a woman to react in ways that reveal something about her personality and suitability as a good match.

In tests, the most highly rated opening lines nearly always start with the man being interested in the woman's surroundings, such as asking about a novel in a bookshop. Such lines may stress intelligence. Humour also works, so long as it isn't cheesy. This is not such a surprise, since laughter triggers the release of feel-good hormones that promote social bonding.

Unsurprisingly, different approaches work for different people. Women who usually have short relationships rather than long ones tend to prefer risqué comments and react better to compliments. Extrovert women appear to prefer humorous lines.

With the ice broken, it's time to think about your style of flirtation. Men who use ostentatious facial expressions, such as raised eyebrows and animated nods of the head, seem to be rated as more attractive by women. This applies even if at the same time they make antisocial comments. The reasoning goes that flirtatious men show social confidence and vigour, which are desirable qualities linked to good genes.

According to psychologists, there are five different flirtation styles. The 'traditional' mode is where a man assumes the conventional gender role, leading the conversation and asking for further dates. Then there's 'sincere', in which both potential partners try to create an emotional bond. 'Playful' style is marked by behaviour that is more fun and superficial. 'Physical' is characterised by flirty body language. While 'polite' describes the most cautious approach.

Be sincere

Of these, the physical and sincere styles work best, leading to many important relationships. The sincere approach yields a greater likelihood of personal and private conversations and romantic interest. Those who adopt the physical style report moving swiftly into relationships.

Whatever your personal style is, before hitting the dance floor or writing an entry on that dating website, it may be worth considering what a potential partner wants from you. And what is your equivalent of the peacock's tail. Once you've got it, flaunt it!

What's in a face?

Dressing up for a date can really pay off. Scientists at Sweden's Uppsala University took pictures of women wearing three different outfits: a dowdy ensemble, their everyday clothes and their glad rags. The women kept their expressions neutral and only their faces appeared in the photos. Asked to rate the attractiveness of the photos, a panel of men consistently chose pictures of the women in their finery, even though the clothes were not visible. It seems that women unconsciously project feelings about their appearance into their facial expressions.

A world of gender identity

For most people gender is a binary affair – a question of male or female. But this limited choice doesn't work for all. Today we know that the range of gender identities – the genders people know themselves to be – is wide and varied. Here are some of the destinations that people travel to beyond the gender assigned to them at birth

Masculine (Masc)
People who identify more with masculinity than femininity

Transsexual man, transsexual male, transsexual person
A transsexual person who identifies as male

Transgender male/man, trans man, trans male, trans dude, trans guy
A person with a male gender identity who was assigned female at birth

Female to male (FTM)
A trans person who identifies as male

Anybody who does not identify with the gender assigned to them at birth and crosses borders to live outside conventional gender identities is a **Trans, transgender, or transgender person**

THE CAPE OF AFAB
A land of people assigned female at birth

Transsexual people undergo hormone therapy or surgery to align their body to their gender identity

Cisgender describes you if you identify with the gender you were assigned at birth

Cisgender female, cisgender woman, cis female, cis woman
A person classed as female at birth who identifies as a female

The distinction between transsexual and transgender is fluid and changes over time

Transgender female/woman, trans woman, trans female, trans girl, trans lady
A person with a female gender identity who was assigned male at birth

AMAB ISLAND
A land of people assigned male at birth

SEA OF GENDER NON-CONFORMITY

Feminine (Femme)
Anyone who identifies more with femininity than masculinity

Transsexual female, transsexual woman, transsexual person
A transsexual person who identifies as female

Male to female (MTF)
A trans person who identifies as female

Cisgender male, cisgender man, cis male, cis man

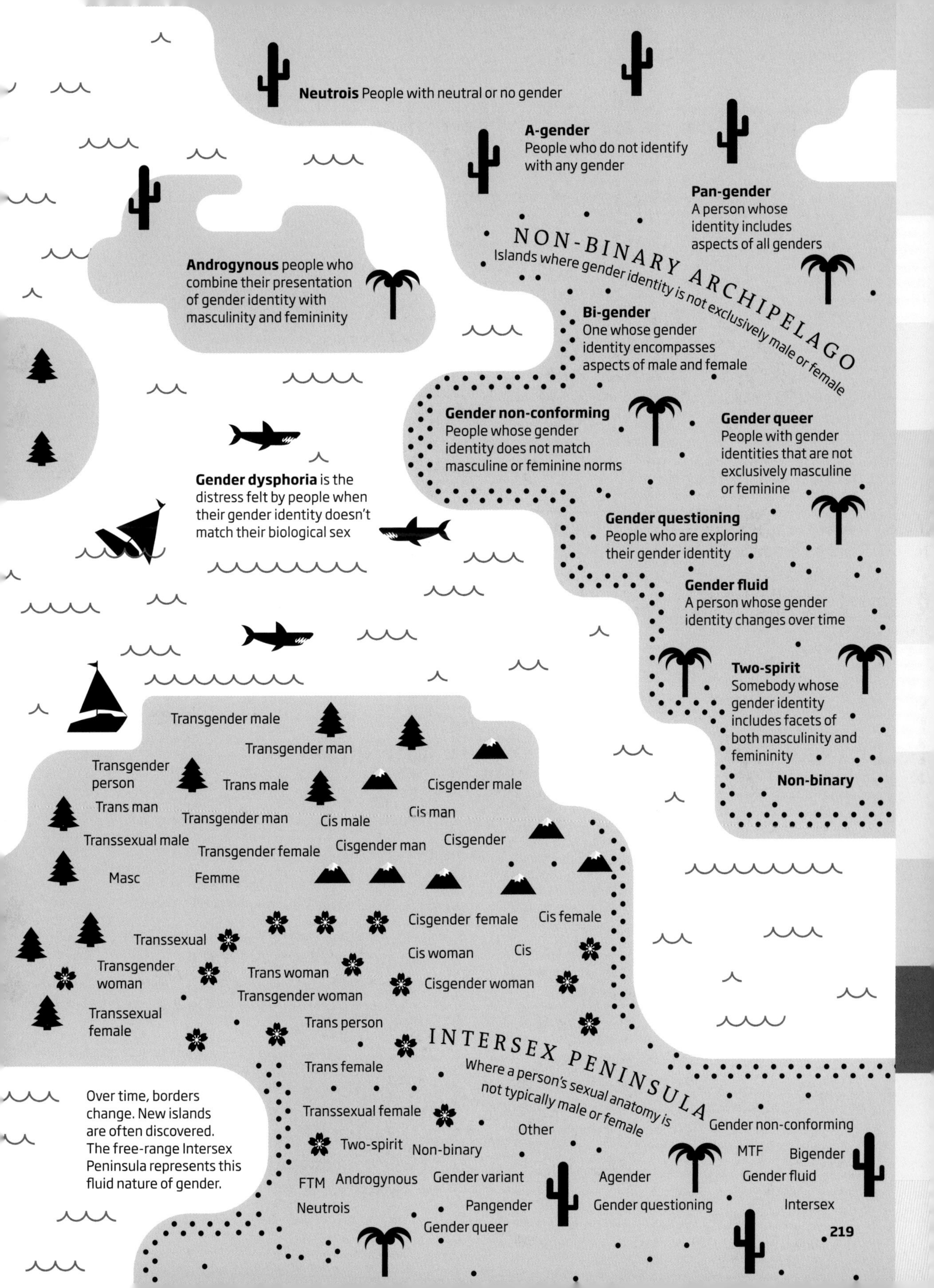

Neutrois People with neutral or no gender
A-gender
People who do not identify with any gender
Pan-gender
A person whose identity includes aspects of all genders
Androgynous people who combine their presentation of gender identity with masculinity and femininity
NON-BINARY ARCHIPELAGO
Islands where gender identity is not exclusively male or female
Bi-gender
One whose gender identity encompasses aspects of male and female
Gender non-conforming
People whose gender identity does not match masculine or feminine norms
Gender queer
People with gender identities that are not exclusively masculine or feminine
Gender dysphoria is the distress felt by people when their gender identity doesn't match their biological sex
Gender questioning
People who are exploring their gender identity
Gender fluid
A person whose gender identity changes over time
Two-spirit
Somebody whose gender identity includes facets of both masculinity and femininity
Non-binary
Transgender male
Transgender man
Transgender person
Trans male
Cisgender male
Trans man
Transgender man
Cis male
Cis man
Transsexual male
Transgender female
Cisgender man
Cisgender
Masc
Femme
Cisgender female
Cis female
Transsexual
Cis woman
Cis
Transgender woman
Trans woman
Cisgender woman
Transgender woman
Transsexual female
Trans person
INTERSEX PENINSULA
Where a person's sexual anatomy is not typically male or female
Trans female
Transsexual female
Gender non-conforming
Other
Two-spirit
Non-binary
MTF
Bigender
FTM
Androgynous
Gender variant
Agender
Gender fluid
Neutrois
Pangender
Gender questioning
Intersex
Gender queer
Over time, borders change. New islands are often discovered. The free-range Intersex Peninsula represents this fluid nature of gender.

Why do we differ so much about sex?

Are you chaste or promiscuous? Would you prefer a short-term fling or a committed relationship? If you're married, do you fantasise about having extramarital sex? Have you ever had more than one sexual partner at a time?

When it comes to sexual attitudes, desires and behaviours, humans are incredibly varied. Why this should be has long intrigued scientists: what makes one person sexually restrained and another uninhibited? Are our beliefs and behaviours predicted by our evolutionary past, or are they individual choices? What role do biology and culture have to play?

Sociosexuality

To discover how sexually unrestricted a person is, the main measure is 'sociosexuality', a score calculated from the answers to a series of intimate questions. These range from the numbers of sexual partners and one-night stands you have had, to whether you fantasise about having sex with someone else while in a relationship.

Such questionnaires reveal that certain attitudes and behaviours go together. For example, people who tend to have more sexual partners are also likely to engage in sex earlier in a relationship, are more likely to have had two or more sexual partners at a time and tend to have relationships characterised by low levels of investment, commitment, love and dependency.

In keeping with popular wisdom, men score high on the sociosexuality scale more often than women. Evolutionary biologists say this is because women bear the heavy costs of pregnancy, breastfeeding and childcare, so it pays them to choose sexual partners carefully or they might get left holding the baby. Evolutionarily, men have not been so tied to child-rearing and have achieved greater reproductive success through multiple short-term relationships. Modern man seems not to be so different.

That is just a first approximation, however. There is a huge overlap in the sociosexuality scores of men and women, and more variation within the sexes than between them. And there are also numerous subtleties in the scores. For example, women's interest in casual sex can change hugely over time. They are more likely to fancy a fling around the time they are ovulating – although they are probably not conscious of it.

At this time, women's preferences can also shift to men who look more masculine and symmetrical,

Early influences

Sociosexuality may be affected by childhood experiences. There is evidence that a stressful home life – perhaps an absent father or marital conflict – can lead girls in particular to breed earlier and more often. The rationale is that there is no point in waiting for a good long-term relationship. Men, too, may be influenced by upbringing. Those who dismiss attachment – who think they are important and others are not worthy of investment and trust – tend to have higher sociosexuality. Such insecurity is thought to arise from stress during childhood caused by unresponsive or erratic caregivers.

which are both indicators of good genes. One suggestion is that women could employ a dual strategy here: a steady mate for child-rearing duties and a fling to obtain some good genes when she is most fertile.

Age is also a factor. Men's sociosexuality peaks in their late twenties. But women are most likely to be unfaithful to their partners when in their early thirties. This is exactly the age when women's chances of conceiving take a tumble and the chances of having a child with a birth defect or genetic problem increases. Could it be that women's increased sociosexuality at this time reflects an evolved strategy to maximise the chances of a healthy child?

So there may be times in any woman's life when being sexually unrestricted is the best strategy. But what about individual differences? Why do some women engage in more casual sex than others, and why does sociosexuality vary so widely in men?

Promiscuous men

One factor is personality. The archetypal promiscuous man scores high on extroversion, low on neuroticism and fairly low on agreeableness. High extroversion gives a man the desire to 'join the chase', low neuroticism means he won't worry about how his behaviour makes him come across, and low agreeableness means he doesn't much care about the social consequences of his actions. Things are similar for women too, though 'openness' – perhaps a desire to try new relationships – also seems to count.

Testosterone levels also play a role. Happily married men and fathers appear to have lower levels of testosterone than other men, while married men on the hunt for extramarital sex have higher levels. This raises the question of whether men with lower than average testosterone are more likely to enter into a committed relationship, or whether being in such a relationship lowers men's testosterone.

A question of looks

Testosterone seems to influence sociosexuality in another way. There is evidence that high testosterone confers a masculine appearance, and we know that these men are especially attractive to women on the lookout for short-term relationships. Could it be that such men are sexually unrestricted because they have more opportunity to be so?

Attractive women, meanwhile, might play the same game, using their looks to attract a partner who has good genes and is faithful. Or, like attractive men, they might simply make the most of their increased opportunities for sex and play the field. The jury is still out on that one. However, one intriguing possibility is that the link between attractiveness and sociosexuality in women can go up and down with the political mood, a bit like hemlines and stock markets. As societies become more liberal and equal, women might be expected to express a greater desire for short-term flings. This is what researchers have found in Scandinavian countries, where attitudes to sex are most liberal.

Status certainly seems to affect a woman's choice of sexual partner. Women who have more control over their finances tend to place higher importance on physical attractiveness in a man than on his financial prospects. Men be warned. If increasing female economic power leads to greater demands for good-looking sexual partners, men may need to invest more in their appearance.

11 Well-being

Food advice to be taken with a pinch of salt

We are constantly being bombarded with health advice, but not all of it is based on rigorous evidence. So which bits can you safely ignore?

Almost everyone 'knows' that we should drink eight glasses of water per day. But that is based on no scientific data whatsoever. The magic number might stem from a 1945 US recommendation that adults consume 1 millilitre of water for every calorie of food, which adds up to about 2.5 litres per day for men and 2 litres for women. Eight 8-ounce glasses of water give you 1.9 litres.

What most people don't realise, though, is that we get much of that water from food. And although, according to the myth, caffeinated drinks don't count because they stimulate the body to lose water, studies don't back this up.

And as for the idea that when you are thirsty you are already dehydrated, that also isn't true. The body is officially dehydrated when blood concentration rises by 5 per cent. We feel thirsty after just a 2 per cent rise. So just relax and drink whatever you fancy whenever you're thirsty.

Time to detox?

A claim with similar traction is that our bodies can and should be detoxed. This idea has intuitive appeal. The modern world is a veritable cesspit of suspect chemicals, many of which we take in with food, water and air. The good news is that you don't have to do anything special to get rid of them. Our livers, kidneys and digestive systems are detoxing all the time, and most of the toxic chemicals we consume are broken down or excreted within hours. However, some substances, especially fat-soluble chemicals such as dioxins and PCBs, can take months or years to get rid of. And if we take these in faster than our bodies can get rid of them, levels start to build up.

Many 'detoxes' advocate dealing with this by fasting or switching to a liquid diet to help the body clean up. But this may well do more harm than good. Fasting or dieting releases fat-soluble chemicals into the blood, which increases their levels in tissues like the muscles and brain, where they could do harm. And there's no guarantee that chemicals released from fat will actually leave the body – some will end up back in fat storage when you go back to eating normally. And when you do, toxin levels will go right back to where they were.

A key part of many people's detox regime is to take antioxidants. Again, the logic seems sound. If free radicals damage cells, and the antioxidants found in fruit and vegetables mop them up, then perhaps popping antioxidant pills is a good way to stay healthy?

Clinical trials beg to differ. While some popular supplements, including beta-carotene, vitamin E and vitamin C, definitely work as antioxidants in the test tube, popping them in pill form does not provide any health benefit. Some studies even suggest that they can be harmful. Supplements of beta-carotene and vitamins A and E, for instance, have been linked to increased mortality.

This could be because the body's own antioxidants are more effective than what we get from food or supplements. So by popping pills we may be overriding a first-rate defence mechanism with a poorer one. One idea is that vegetables are beneficial because they are mildly poisonous, and activate protective mechanisms that ward off disease. If this is right, the benefits of vegetables may have nothing to do with antioxidants after all.

That surely adds support to the claim that we should eat and live like cavemen. Our bodies evolved to hunt game and gather fruit and vegetables. So perhaps we'd all be a lot healthier if we lived and ate more like our ancestors – eating game, fish, fruit, non-starchy vegetables and nuts.

Some of this makes good nutritional sense. Other aspects of the 'paleo' diet, such as ditching grains, legumes and dairy, do not. The idea touted by such diets, that we haven't adapted to our agricultural diet, actually isn't true: many people have extra copies of genes for digesting the starch found in grains. The ability to digest milk as an adult – lactose tolerance – has also evolved independently in several populations.

Clouded by ignorance

Apart from that, we don't know for sure what our ancestors ate: their diets probably varied from place to place. And there is no reason to think that our ancient ancestors hit an evolutionary 'sweet spot' when they were perfectly in tune with their environment. We really have no idea whether they were healthier than us or not.

But our ancestors certainly weren't fat. So does that mean that we should always strive not to get overweight?

Being seriously obese is very bad for your health. Carrying just a few extra pounds, though, may actually deter the grim reaper. According to analyses of large numbers of people, being 'overweight' – defined as having a body mass index (BMI) of 25 to 29 – brings a 6 per cent reduction in death risk compared with people with a BMI of between 18.5 and 25.

Why, isn't clear. Perhaps carrying a few extra pounds in reserve helps the body fight off illness or infection. Perhaps overweight people are more likely to receive medical attention. Or perhaps some of those counted as 'normal' were actually ill. Whatever the reason, it seems that a little bit of flab may not be a crime against health after all.

Does sugar make children hyperactive?

Many parents will find it hard to believe, but sugar does not cause hyperactivity. Double-blind studies, where no one at the time knew which kids had received sugar and which a placebo, found no differences in the children's behaviour. Parents, though, spot 'hyperactive' behaviour in their offspring after being told the kids have been given sugar, even when, in reality, they had not.

If anything, sugar affects kids' brains in exactly the opposite way, helping them to concentrate on tasks and score better on memory tests. Whereas an inability to concentrate is a characteristic of hyperactivity. So perhaps what parents mistake for hyperactivity at parties is just sugar-fuelled kids concentrating on having fun.

Caveat emptor

Superfoods and supplements improve your health, right? Not necessarily. It appears the more companies advertise them, the less likely they are to do you good

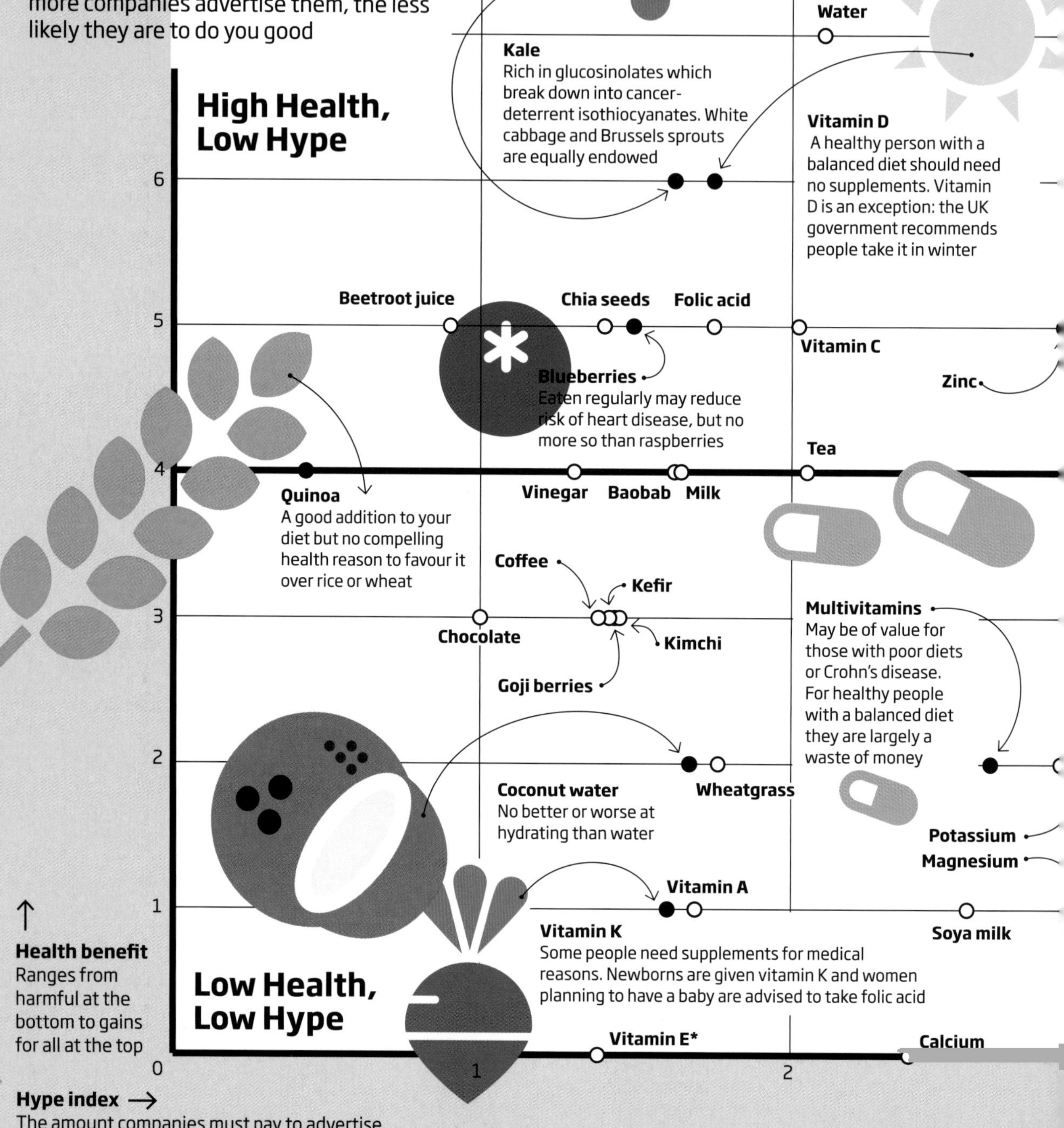

↑

Health benefit
Ranges from harmful at the bottom to gains for all at the top

Hype index →
The amount companies must pay to advertise products online gives a measure of the competition - the hype - surrounding them

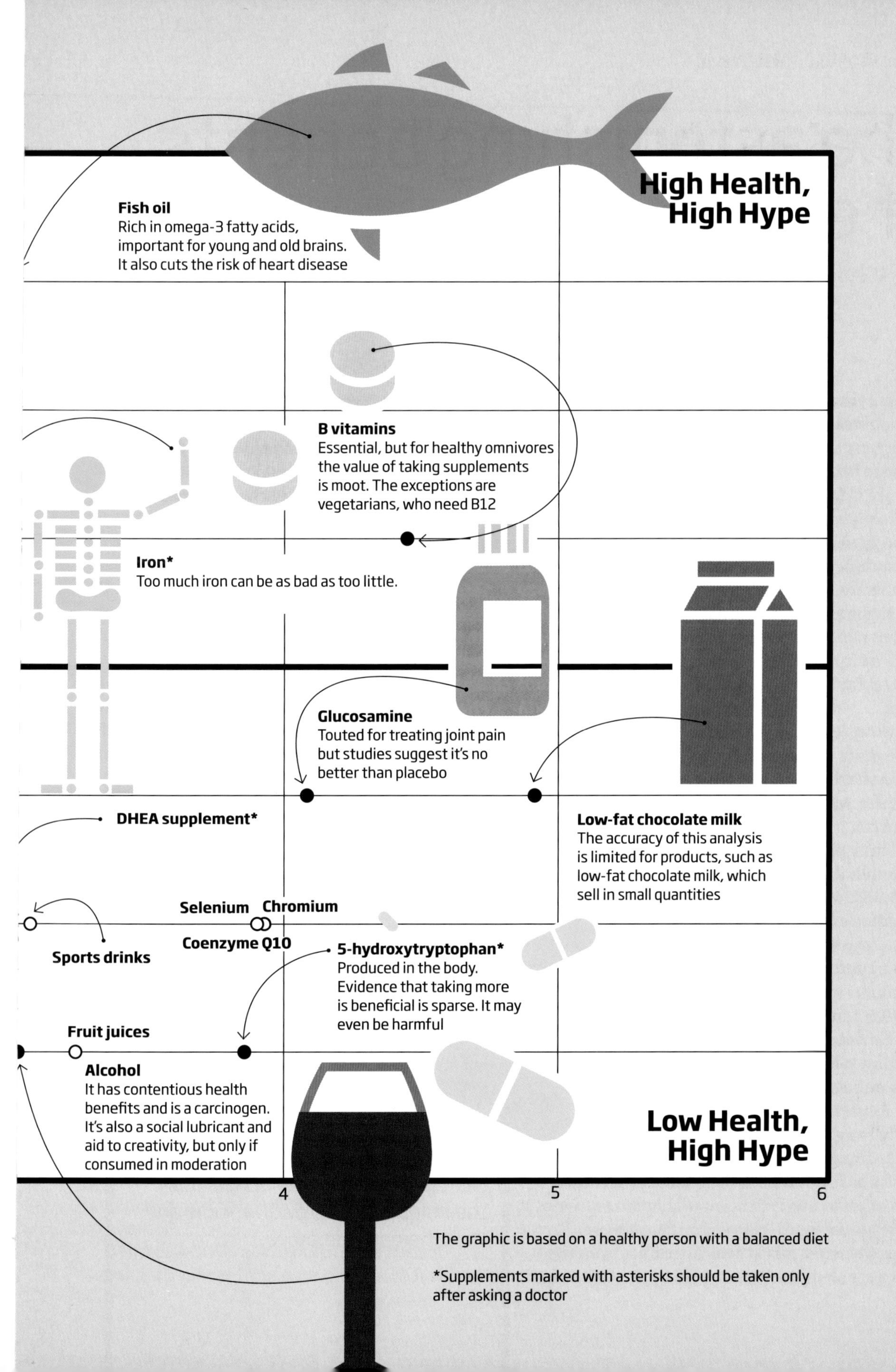

The graphic is based on a healthy person with a balanced diet

*Supplements marked with asterisks should be taken only after asking a doctor

Faster body, faster mind

If you fancy a bit of brain training, drop the puzzle pencil and get your running shoes on.

We all know that keeping fit is good for our bodies. Lack of exercise opens the door to everything from obesity to type 2 diabetes and heart disease (see 'The astonishing benefits of exercise', p 228). In comparison, the effects of physical activity on our brains have been undersold.

Yet research now shows that physical fitness isn't just a 'nice to have' for our mental powers. It has a profound, long-lasting impact on a range of cognitive abilities. It's important throughout life, for children who want good exam results right up to elderly people who are keen to avoid a slow slide into senility.

Hard evidence of a link between cognitive performance and fitness emerged only in the 1990s. In animals, research showed that exercise stimulates the growth of new neurons in mice. In people, it found that sedentary adults who exercise aerobically over several months improve their ability in cognitive exercises that need executive control. That's the kind of control that enables you to change between different tasks without making errors. It's a key ingredient of greater general intelligence.

Since then, more evidence has stacked up, especially for elderly people. Over-fifty-fives who do not exercise tend to have poorer cognitive functions – such as memory and learning skills – than contemporaries who swim, garden or cycle a few times a week. Such activities also appear to have lasting benefits: middle-aged people who exercise at least twice a week cut their risk of developing dementia in their sixties and seventies.

For other age groups, there are fewer research results. But what exists supports the notion that physical activity benefits the brain. For example, schoolchildren up to fourteen years of age who are in the peak of fitness tend to do much better in standard academic tests than those who are rated as most unfit. And in young men, cardiovascular fitness correlates well with intelligence and is a good predictor of educational achievement later in life.

Tuning up the brain

What lies behind this long-lasting effect? The brain needs a large supply of oxygen and nutrients via an intricate network of capillaries. Exercise encourages growth of these supply lines. Also, high blood pressure in the large, central arteries that feed the brain can reduce cognitive performance, perhaps by disrupting the capillaries. Fit people tend to have lower blood pressure, which should help to protect against such damage. Being fit also cuts the risk of developing type 2 diabetes, a condition that increases the risk of developing Alzheimer's disease. So staying fit should stave off dementia.

Exercise keeps the brain running smoothly, too. It stimulates the release of neurotransmitters, such as dopamine, serotonin and noradrenaline, which regulate signalling in the brain. These are the same compounds that drugs for attention-deficit hyperactivity disorder and depression are designed to increase. Physical activity also triggers the release of compounds, such as 'insulin-like growth factor-1' and 'brain-derived neurotrophic factor', that encourage the growth of new neurons and connections between them.

The origins of the link between body and brain lie in our evolutionary past and are still largely obscure. Yet they could lie behind a profound advance in our development. A distinguishing factor of our ancestors is that they became adept at running down prey over long distances. In doing

Work that brain

Aerobic exercise is essential for your brain but it doesn't need to be strenuous: walking briskly a few times a week can work well. But if you're already fit, high-intensity interval training (HIT) can take it to the next level. It consists of short, very hard bursts of exercise which stimulate the pituitary gland to release human growth hormone which, in turn, boosts levels of neurotransmitters. HIT can also boost 'brain-derived neurotrophic factor', which encourages the growth of new brain cells, and pushes cognitive skills higher than more leisurely activities.

so they would have experienced a constant flow of those neuron-boosting chemicals. Could it be that increased endurance running triggered a leap in intelligence?

Competitive edge

If greater athletes became smarter – and hence better – hunters, they would have had a competitive edge, not least in the mating game. Alternatively, those chemicals may have merely boosted hunters' physical endurance, and greater brainpower was a by-product of this – an evolutionary accident.

However it happened, evolutionary links between athletic and mental prowess are supported by a variety of studies. One measured the brains and exercise capacity of groups of animal species, such as rodents, dogs and cats. Within each group, the species with the largest capacity for physical activity tended to have the biggest brain in relation to its total body mass. Another study found that mice selectively bred for long-distance running have increased production of new cells in the hippocampus and striking growth in other brain regions.

Fossil evidence from our ancestors also shows that brain size appears to have increased in step with traits that indicate greater capacity for physical activity, such as longer limbs. However, while the link between greater athleticism and smarter brains is firming up, we still cannot say that one caused the other.

Tremendous impact

Whatever role physical activity played in our past, its importance today is potentially huge. The US government recommends that children from six to eighteen years of age should do at least 60 minutes a day of aerobic exercise, because it keeps not only their bodies in peak condition but also their brains. That reasoning applies just as well to adults. New exercise regimes that typically last for six months or more tend to increase brain processing speed and improve memory and attention.

In years gone by, scientists thought that physical activity enhanced a healthy brain. That view has now changed. Today the thinking is that a healthy brain is impossible without a highly active body. In other words, exercise is not an enhancer of normal cognition, it is a necessity. Whether you are eight or eighty years old, the message is clear: get a move on!

The disturbing truth about sitting down

Are you sitting comfortably? Then DON'T. Even if you're fit and active, sitting on your backside is seriously bad for your health. Every hour spent slumped in front of the television slices about 20 minutes off your life expectancy. Sitting in front of a computer at work is no better. Even curling up and reading a book – surely a blameless activity – can do terrible things to your metabolism.

That may sound like a statement of the obvious: we have the expression 'couch potato' for a reason. But the killer point is this: sitting on your butt is bad for you even if you exercise as well. Heading to the gym or going for a run is not a licence to spend the rest of the day on your backside. Just as you cannot compensate for smoking 20 a day by running 10 kilometres at the weekend, a bout of exercise does not cancel out the effect of watching TV for hours on end.

People who spend hours sitting have a higher mortality rate even if they work out for 45 to 60 minutes a day. Researchers call these people 'active couch potatoes'.

Dying young

The effect of sedentarism is very stark. If you compare middle-aged adults who spend 6 hours a day sitting down with those who spend just 3 hours – taking into account other health risks such as diet and smoking – the couch potatoes have a mortality rate 27 per cent higher. Exercise has no impact on this figure. Or to put it another way, somebody who spends 6 hours a day watching television can expect to die 5 years younger than somebody who doesn't.

But it is not just the couch that should worry people. The harm comes primarily through inactivity itself, and so other kinds of inactivity may be just as harmful, be it sitting in a car or at an office desk. (Sleep brings its own health benefits and probably does not count, though people who sleep more than 9 hours a day are at greater risk of dying, which may have something to do with extended inactivity.)

Occupational hazard

Unsurprisingly, people do a lot of these kinds of things. Most people spend more than half of their 14 to 15 waking hours sitting down. Moderate-to-vigorous activity – aka 'exercise' – occupies just 5 per cent or less of our days.

For a lot of people, parking the arse is quite literally an occupational hazard. For somebody with a desk job, a typical working day might involve sitting for 7 to 8 hours, often 2 to 3 hours at a stretch.

Of course many jobs entail very little sitting at all – hairdressers, construction workers, chefs, waiters, nurses and many others are typically on their feet most of the working day. But such jobs are becoming less common, and people who do them are just as inclined as anyone to collapse onto the sofa once the day is done. That's not the lifestyle to which the human body is adapted. From an evolutionary point of view, we are built to be active.

Metabolic cascade

From studies of extended inactivity, we know that our ill-adapted bodies respond with a complex cascade of metabolic changes. For example, unused muscles shift from burning fat to burning glucose. With them relying more on carbohydrates for what little work they are doing, unburned lipids accumulate in the blood, which could be why sitting has been linked to heart disease. Fat also gathers in muscles, the liver and the colon – places where it is not supposed to be stored.

Other changes involve insulin resistance, a diabetes-like condition in which glucose accumulates in the bloodstream even when the body produces enough insulin to deal with it.

So what can people do to avoid these detrimental effects, other than quitting their desk jobs and taking up nursing, hairdressing or waiting on tables? First of all, it is important to realise that exercise still has benefits – an hour's workout may not undo hours of being a couch potato, but it is still good for your health in other ways. Extended periods of sitting down should be seen as an independent risk factor that should be addressed separately.

Frequent bursts

But how? The rule of thumb is that you need to break up your sitting with brief but frequent bursts of light activity. That means raising your metabolic rate to just 1.5 times its resting rate, which is actually very easy to do: just standing up and walking around is more than enough. Laboratory studies suggest that doing this for two minutes every 20 minutes is sufficient to wipe out the negative effects of extended sloth.

What makes these short bouts of activity effective is that they're enough to burn off some of the glucose that's accumulated in your bloodstream. Given that a healthy blood sugar level is about 1 gram of glucose per litre of blood, you don't have to burn very much glucose to get your level down. If you burn off just 4 calories by, say, stacking or unstacking the dishwasher or ambling to the water fountain, that's a gram of glucose gone.

This may sound like a chore but is actually good news for the millions of people who have not been able to get close to recommended daily exercise levels. Anything is better than nothing. Just getting up and moving at all is taking a step in the right direction.

Those glucose-burning chores

If standing up and ambling about to break up your loafing habit feels too aimless, there are plenty of gentle household chores that will do the job. Loading clothes into a washing machine, doing the dishes and showering all count as physical activity, albeit of the lowest intensity. Walking up a flight of stairs, sweeping and making the bed are classed as 'moderate activity'. And if you fancy a snack to take back to the couch, so is preparing food. Unfortunately, yawning, stretching, chewing, opening a beer and scratching your rump are not vigorous enough.

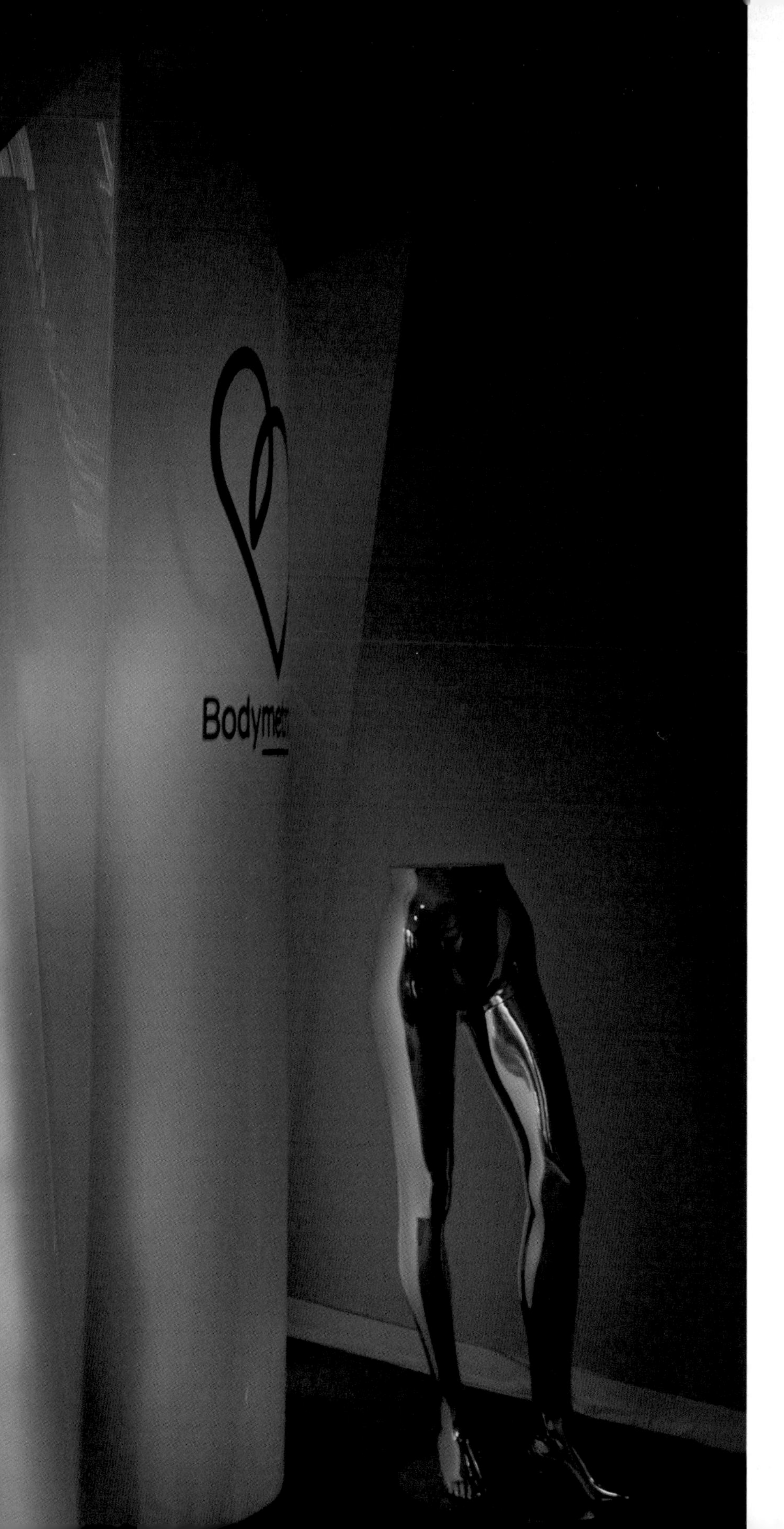

Does my belly look big in this?

It seems *de rigueur* to check how many steps you take each day and to keep an exercise diary. If you're really keen, some apps will keep track of what you've drunk and eaten so you can monitor your energy balance. For some people, though, that isn't enough. Yet there's a level above even this for monitoring the impact of exercise and diet. Suran, left, is one of these extreme life loggers. He regularly steps into a digital 3D body scanner to record his vital statistics. After his uncle died of a heart attack, he decided to keep an eye on his risk of heart disease, a good predictor of which is waist size. A tape measure proved too inaccurate so he opted for the scanner. More often found in clothing stores, these scanners not only stop you committing fashion crimes, they may even save your life.

Credit: Travis Hodges/ INSTITUTE

How to think yourself healthy

There is a drug that we can all get for free that has wide-ranging health benefits and no side effects. It's called the mind. From the placebo effect to hypnosis, knowing how to harness its power can do you the power of good.

One of the mind–body connections we know most about is the placebo effect. If you regularly take medication, try this trick. Before you swallow your pills give them a pep talk. Something like, 'Hey guys, I know you're going to do a good job'.

That might sound eccentric, but based on what we've learned about the placebo effect, there is good reason to think that talking to your pills really can make them do a terrific job. The way we think and feel about medical treatments can dramatically influence how our bodies respond.

Simply believing that a treatment will work may trigger the desired effect even if the treatment is inert – a sugar pill, say, or a saline injection. For a wide range of conditions, from depression to Parkinson's, osteoarthritis and multiple sclerosis, it is clear that the placebo response is far from imaginary. Trials have shown measurable changes such as the release of natural painkillers, altered neuronal firing patterns, lowered blood pressure or heart rate and boosted immune response, all depending on purely psychological cues. There is even evidence that some drugs work by amplifying a placebo effect – when people are not aware that they have been given the drugs, they stop working.

Not just a con trick

It has always been assumed that the placebo effect only works if people are conned into believing that they are getting an active drug. But now it seems this may not be true. Belief in the placebo effect itself – rather than a particular drug – might be enough to encourage our bodies to heal. A team of doctors at Harvard Medical School in Boston gave people with irritable bowel syndrome an inert pill. They told the patients that such pills have been shown in clinical studies to significantly improve IBS symptoms through mind–body self-healing processes, which is perfectly true.

Despite knowing the pills had no active ingredient, on average the volunteers rated their symptoms as moderately improved after taking them, whereas those given no pills said there was only a slight change. Trials have since found

Do what you love

Many researchers who work on the mind-body connection think what really matters is having a sense of purpose in life. Having an idea of why we are here and what is important increases our sense of control over events, rendering them less stressful. One study of a three-month meditation retreat found that the physiological benefits correlated with an increased sense of purpose in life. The participants were already keen meditators, so the study gave them lots of time to do something important to them. Simply doing what you love, whether it's gardening or voluntary work, might have a similar effect on health.

Jo Marchant is a London-based science journalist and author of *CURE: A Journey into the Science of Mind over Body*

benefits for such 'honest placebos' in depression, migraine, back pain and ADHD. These studies raise the possibility that we could all use the placebo effect to convince ourselves that something simple such as sucking on a sweet or downing a glass of water will banish a headache, clear up a skin condition or boost the effectiveness of drugs we take.

Positive thoughts

Another way you can boost your health is to give yourself (rather than your pills) a pep talk. Think positive and tell yourself everything's going to be fine. Again, it sounds too good to be true, but realism can be bad for your health. Optimists recover better from medical procedures such as coronary bypass surgery, have healthier immune systems and live longer.

It is well accepted that negative thoughts and anxiety can make us ill. Stress – the belief that we are at risk – triggers physiological pathways such as the 'fight or flight' response, mediated by the sympathetic nervous system. These have evolved to protect us from danger, but if switched on long-term they increase the risk of chronic conditions from diabetes to dementia.

Positive beliefs help to quell stress, and some researchers believe they have a separate positive effect too – feeling safe and secure, or believing things will turn out fine, encourages the body to put resources into growth and repair (as opposed to defence).

Optimism seems to reduce stress-induced inflammation and levels of stress hormones such as cortisol. It may also reduce susceptibility to disease by dampening asctivity in the sympathetic nervous system and stimulating the parasympathetic nervous system. The latter governs what's called the 'rest and digest' response – the opposite of fight-or-flight.

Just as helpful as taking a rosy view of the future is having a rosy view of yourself. High 'self-enhancers' – people who see themselves in a more positive light than others see them – have lower cardiovascular responses to stress and recover faster, as well as lower baseline cortisol levels.

Whatever your natural disposition, you can train yourself to think more positively, and it seems that the more stressed or pessimistic you are to begin with, the better it will work.

Your attitude towards others can also have a big effect on your health. Being lonely increases the risk of everything from heart attacks to dementia, depression and death, whereas people who are satisfied with their social lives sleep better, age more slowly and respond better to vaccines. The effect is so strong that curing loneliness is as good for your health as giving up smoking.

Get a (social) life

Put simply, people who have rich social lives and warm, open relationships don't get as sick and they live longer. This is partly because people who are lonely often don't look after themselves well, but there are direct physiological mechanisms too – related to, but not identical to, the effects of stress.

In lonely people, genes involved in cortisol signalling and the inflammatory response are up-regulated, and immune cells important in fighting bacteria are more active. Researchers suggest that our bodies have evolved so that perceiving ourselves to be socially isolated triggers branches of the immune system linked to wound healing and bacterial infection. That makes sense, because an isolated person would be at greater risk of physical trauma.

Crucially, these differences relate most strongly to how lonely people believe themselves to be, rather than to the actual size of their social network. That also makes sense from an evolutionary point of view, because being among hostile strangers can be just as dangerous as being alone. There is evidence that over time, lonely people become overly sensitive to social threats and come to see others as potentially dangerous. Tackling this attitude reduces loneliness more effectively than giving people more opportunities for interaction, or teaching them social skills.

If you feel satisfied with your social life, whether you have one or two close friends or many, there is nothing to worry about. But if you feel alone and threatened by others, it's probably time to take action.

Time to meditate

One option might be to find a new pastime – and why not kill two birds with one stone and enrol in meditation classes?

Monks have been meditating on mountaintops for millennia, hoping to gain spiritual enlightenment. Their efforts have probably enhanced their physical health, too.

Trials looking at the effects of mindfulness meditation suggest that it reduces physical symptoms such as pain, anxiety, depression and fatigue. There is also some evidence that meditation boosts the immune response in vaccine recipients and people with cancer, soothes skin conditions and even slows the progression of HIV.

Meditation might even slow the ageing process. Telomeres, the protective caps on the ends of chromosomes, get shorter every time a cell divides and so play a role in ageing. In one study, three months of meditation increased levels of an enzyme that builds up telomeres. As with social interaction, meditation probably works largely by influencing stress-response pathways. People who meditate have lower cortisol levels, and they have changes in their amygdalae, brain structures important in fear and the response to threat.

Mesmerising

Another old-time method gaining in respectability is hypnotherapy, invented by Franz Mesmer in the eighteenth century.

Peter Whorwell of the University of Manchester has spent much of his professional life building a body of evidence for the use of hypnosis to treat just one condition: irritable bowel syndrome. IBS is considered a 'functional' disorder – a rather derogatory term used when a patient suffers symptoms but doctors can't see anything wrong. Whorwell felt that his patients, some of whom had such severe symptoms they were suicidal, were being let down by the medical profession. That's why he got into hypnosis.

Whorwell gives patients a brief tutorial on how the gut functions, then gets them to use visual or tactile sensations – the feeling of warmth, for example – to imagine their bowel working normally. It seems to work: IBS is the only condition for which hypnosis is recognised as a possible treatment by the UK's National Institute for Health and Care Excellence, though only for those who have failed to respond to other treatments.

It isn't clear how hypnosis works. But when hypnotised, people can influence parts of their body in novel ways. Under hypnosis, some IBS patients can reduce the contractions of their bowel, something not normally under conscious control. Their bowel lining also becomes less sensitive to pain.

Hypnosis may tap into physiological pathways similar to those involved in the placebo effect. For one thing, the two benefit similar medical conditions, and both are underpinned by suggestion and expectation.

Most clinical trials involving hypnosis are small, but they suggest that hypnosis may help pain management, anxiety, depression, sleep disorders, obesity, asthma and skin conditions such as psoriasis and warts. Hypnotising yourself seems to work just as well. In fact, Whorwell thinks self-hypnosis is the most important part.

Find a purpose

If none of these work, you can always try to find God. In a study of 50 people with advanced lung cancer, those judged by their doctors to have high 'spiritual faith' responded better to chemotherapy and survived longer. More than 40 per cent were still alive after three years, compared with less than 10 per cent of those judged to have little faith. Are your hackles rising? Of all the research into the healing potential of thoughts and beliefs, studies into the effects of religion are the most controversial.

There are thousands of studies purporting to show a link between some aspect of religion – such as attending church or praying – and better health. Religion has been associated with everything from lower rates of cardiovascular disease to better immune functioning and improved outcomes for infections such as HIV.

Critics of these studies point out that many of them don't adequately tease out other factors. For instance, religious people often have low-risk lifestyles, churchgoers tend to enjoy strong social support, and seriously ill people are less likely to attend church. Nonetheless, one analysis concluded, after trying to control for these factors, that 'religiosity/spirituality' does have a protective effect, though only in healthy people. The authors warned there might be a publication bias, though, with researchers failing to publish negative results.

Even if the link between religion and better health is genuine, there is no need to invoke divine intervention to explain it. Some researchers think the positive emotions associated with spirituality promote beneficial physiological responses. Others attribute it to the placebo effect – trusting that some deity or other will heal you may be just as effective as belief in a drug or doctor. Like a sugar pill, God doesn't need to be real to make you better.

Take a moment

If you fancy trying the health benefits of meditation but don't have time for a three-month retreat, don't worry. Imaging studies show that the positive structural changes in the brain associated with meditation can be gained after as little as 11 hours of training. It is even possible to do short 'mini-meditations' throughout the day, taking a few minutes at your desk to focus on your breathing, for example. According to the experts, little moments here and there all count.

How to shake off that feeling of fatigue

You're in bed by 10.30 after a busy, productive day. After a full night's sleep you wake up naturally and feel ... exhausted.

If this sounds familiar, you're not alone. Complaints about tiredness account for around a third of visits to the doctor. Workers who turn up too tired to work effectively cost US employers an estimated $100 billion a year. So it is perhaps surprising that scientists are only now beginning to investigate what fatigue actually is.

Fatigue versus sleepiness

Until recently, tiredness during the day was usually blamed on inadequate sleep. The US Centers for Disease Control and Prevention estimates that 35 per cent of people are short on sleep. Yet sleep researchers consider sleepiness and fatigue to be different things.

The sleep latency test, widely used in sleep clinics, is one way to tell the difference. The idea is that if you lie down somewhere quiet during the day and fall asleep within a few minutes, then you are either lacking sleep or potentially suffering from a sleep disorder. If you don't drop off within around 15 minutes, yet still feel tired, fatigue might be the problem.

So if fatigue isn't sleepiness, what is it? One idea is that it stems from a problem with the circadian clock, which regulates periods of mental alertness through the day and night. This regulation falls to the brain's 'suprachiasmatic nucleus' (SCN), which coordinates hormones and brain activity to ensure that we feel generally alert by day. Under normal circumstances, the SCN orchestrates a peak in alertness at the start of the day, a dip in the early afternoon, and a shift to sleepiness in the evening.

The amount of sleep you get at night has little impact on this cycle. Instead, alertness depends on the quality of the hormonal and electrical output signals from the SCN. The SCN sets its clock by the amount of light hitting the retina, so that it keeps in line with the solar day. Too little light in the mornings, or too much at night, can disrupt SCN signals, and either can lead to a lethargic day.

Cutting body fat

Exercise also seems to reset the SCN, plus it has the benefit of reducing body fat, which also plays a role in fatigue. Cutting body fat may help to reduce fatigue. Body fat not only takes more energy to carry around, but releases leptin, a hormone that signals to the brain that the body has adequate energy stores. Studies have linked higher leptin levels to greater perceived fatigue, a finding that makes perfect sense from an evolutionary perspective: if you aren't short of food, you don't need the motivation to go out and find some.

People who carry excess fat also show higher levels of inflammation, an immune response that triggers the release of proteins called cytokines into the bloodstream. Body fat stores large quantities of cytokines, which may mean that more end up circulating, too. These cytokines rouse other parts of the immune system into action and put the brain and body into 'sickness mode' where rest and recuperation is all that we have the energy for.

Even if you're not overweight or sick, inflammation could still be running you down. A sedentary lifestyle, regular stress and poor diet – one high in sugar and low in fruit and vegetables – have all been linked to chronic, lower-level inflammation. There is also preliminary evidence that disruption of circadian rhythms can increase inflammation in the brain. It's early days, but it could be that lifestyle-related inflammation is to blame for much of modern tiredness, linking

Causes new and old

One possible cause of widespread fatigue is that life is more exhausting than it has ever been. Caught between the competing demands of work and family, not to mention the ever-present buzz of smartphone notifications, it is no surprise so many of us feel as if we are running on empty. Yet this may be a fallacy. People through the ages have consistently complained of being worn out, and harked back to the relative calm of simpler times. Over the centuries, fatigue has been blamed on the alignment of the planets, a lack of godliness and even an unconscious desire to die

fatigue to everything from poor-quality sleep and physical inactivity to a bad diet.

If that's correct, then a handful of lifestyle changes could go a long way to fighting everyday fatigue. Some studies suggest that inflammation is reduced by taking more exercise, and by eating more fruit and vegetables that contain high levels of polyphenols (such as resveratrol in grapes or curcumin in turmeric). There is some evidence that iron supplements provide an energy boost, even in people without clinical iron-deficient anaemia.

Raising motivation

Another factor muddying the waters is that biological signals that might lead one person to experience overwhelming exhaustion won't necessarily trigger it in another. Some people are able to push through it.

That requires motivation, low levels of which are clearly an important aspect of fatigue. So some researchers have been looking at the role of dopamine – a neurotransmitter that drives us to seek out pleasure. When dopamine levels take a plunge, as happens in Parkinson's disease, for example, the accompanying depression and apathy can be crushing.

Since the vast majority of people with major depression report severe fatigue, and about one in five people become depressed at some point in their lives, it's no wonder that depression is also a potential common factor in fatigue. Indeed, widespread depression may be an explanation for why so many of us feel so drained.

Stress and enjoyment

With so many emerging causes for fatigue, interest in trying to crack the problem is growing. In the meantime, the best advice is not to let fatigue stop you doing something you enjoy. In fact, it is worth forcing yourself to keep at it because a potent reward could trigger the release of dopamine in brain areas linked to motivation and alertness. Alternatively, do something stressful: the release of adrenaline could help you overcome lethargy. Ideally, put stress and enjoyment together. After all, nobody feels fatigued when they're on a roller coaster.

How much sleep do we need?

An obsession with sleep often consumes our waking hours. Sleep is as vital for life as food or water. Lab rats deprived of it die after a few weeks, and people who inherit the rare disease 'fatal familial insomnia' meet the same fate within 18 months of diagnosis. We still don't know what sleep is for, or why total lack of it kills you. But the many ways it affects our well-being are becoming increasingly clear.

Sleep has been labelled the third pillar of good health, along with diet and exercise. But that's underselling it: sleep is the foundation on which the other two pillars rest. Almost all physical and mental processes are boosted by good sleep and impaired by lack of it.

Poor sleep is associated with memory impairment, emotional dysregulation and poor decision making. It affects your immune system and appetite, and has been linked to metabolic diseases such as obesity and type 2 diabetes. Increasingly, lack of sleep is implicated in mental health problems including depression, bipolar disorder and schizophrenia, and neurological conditions such as Alzheimer's disease. It's enough to keep you awake at night.

To make matters worse, there's a general perception that we are collectively sleep-deprived, often as a result of the pressures of modern life (though this is not a recent worry – see box, right). There is some evidence that this is true. If people are asked whether they think they get enough sleep and would they like to get more, they usually answer 'no' and 'yes'. The Royal Society for Public Health says Britons get an hour less than they need each night, while a third of adults experience symptoms of insomnia.

But many sleep researchers think this is nonsense. There is little evidence that most adults do not get enough sleep, and our collective sleep debt, if it exists at all, has not worsened in recent times. So how much sleep do you really need? And what is the best way to get it?

The magic number

We all know 8 hours is the magic number for a decent night's sleep. But nobody seems to know where this number came from. In questionnaires, people tend to say they sleep for between 7 and 9 hours a night, which might explain why 8 has become a rule of thumb. But people also tend to under- or overestimate how long they have been out for the count.

A modern curse

There is a widespread belief that we sleep less than our grandparents did, and that modern society is in the grip of a sleep-deprivation epidemic caused by work pressure and the frenetic pace of life. *Plus ça change*. Claims of widespread sleep deprivation in Western society are nothing new – in 1894, the *British Medical Journal* ran an editorial warning that 'The subject of sleeplessness is once more under public discussion. The hurry and excitement of modern life is quite correctly held to be responsible for much of the insomnia of which we hear.'

Studies of hunter–gatherer societies with no access to electricity suggest that our need for sleep is actually a bit less than 8 hours. Such people typically sleep for 6 or 7 hours without triggering any of the ill effects associated with sleep deprivation. So perhaps 8 hours is the wrong target and we can get by just fine with 7. If anything, this seems to be a minimum requirement. Regularly getting less sleep than that increases the risk of obesity, heart disease, depression and early death.

Even by this lowered benchmark, sleep deprivation is common. Around a third of US adults get less than 7 hours. The average in the UK is 6.8.

But when it comes to sleep, averages are not all that useful. Sleep requirement varies widely from person to person and is largely genetic. It also changes as you age. Taking this into account, the US National Sleep Foundation recommends 7 to 9 hours for adults, but with leeway of an hour either side to account for natural variation.

Rule of thumb

Hmm. That isn't actually much use, suggesting that anything between 6 and 10 hours a night is fine. So how much is enough for you? As a rule of thumb, if you need an alarm clock to wake you in the morning, you are not getting enough. But you can also have too much of a good thing. Regularly getting more than 8 hours increases your chances of dying, and the association is at least as strong, often stronger, than the association of sleep deprivation with mortality.

Just why this is so remains a mystery. It could be that people with underlying health problems sleep more. Or maybe it is simply that when we are asleep we are moving very little, and inactivity is bad for you (see 'The disturbing truth about sitting down', p 232). Long sleep is also associated with inflammation, an immune response linked to everything from depression to heart disease.

What of those annoying people who claim to get by just fine on a few hours each night? They probably are sleep-deprived, but have got used to the effects and now fail to notice them. Only a tiny minority of us, probably less than 3 per cent, can get by on 6 hours or less with no problems at all.

Regardless of how much sleep we need, sleep deprivation is often a fact of life. For many people the need to work plus the temptations of social life or home entertainment lead to candles being burned at both ends.

So what happens when we stay up too late, or get up too early, or both? The need to sleep is controlled by a two-tier system. The circadian clock relies on light to keep your sleep/wake pattern within around 24 hours. Then there's sleep drive or sleep pressure. The longer you are awake, the more a chemical called adenosine builds up in your brain, sending signals that increase your desire for sleep. After prolonged wakefulness – 16 hours or more – the pressure becomes overwhelming and you have to fight the urge to drop off.

If you override the pressure – say by drinking caffeine, which blocks adenosine receptors in the brain – the effects quickly show themselves. Being awake for 24 hours will leave you with the same level of cognitive impairment as having a blood alcohol content of 0.1 per cent – more than the drink-drive limit in many countries.

Chronic lack of sleep takes a toll, too. Sleeping for just 4 hours a night for several nights on the trot quickly leads to high blood pressure, increased levels of the stress hormone cortisol and insulin resistance – a precursor to type 2 diabetes. It also suppresses the immune system.

Fortunately, these acute effects can be reversed by repaying your sleep debt with an early night or a lie-in – which is what many people do at the weekend.

Repaying your sleep debt

When a long sleep is impossible, napping can compensate for sleep debt. Once a sign of laziness, it's now clear that taking 40 winks is a great way to improve your performance.

A 'nano-nap', lasting just 10 minutes, can boost alertness, concentration and attention for as much as 4 hours. Double it and you increase your powers of memory too. Either way, you are unlikely to enter the deeper stages of sleep, so will avoid the phenomenon known as sleep inertia, the groggy feeling that can occur when waking from deep sleep. On the flip side, you won't get the benefits of deep sleep, which provides the biggest boost to learning. If that's your aim, opt for a nap of between 60 and 90 minutes, which gives you time to go through a full cycle of sleep. This aids learning by shifting memories from short-term storage to lockdown in longer-term memory vaults.

If you feel in need of a nap, it's easy. Find a warm, dim and quiet place to lie down. And if you want to keep it short, drink a cup of coffee immediately beforehand – the caffeine kicks in after about 20 minutes, snapping you back to wakefulness without sleep inertia.

However, going for a nap requires you to recognise you need more sleep, and the more sleep-deprived you become, the more you underestimate how tired you are. If you have chronic sleep loss – after a hectic period of work, for example – only a proper holiday can break the cycle.

There's a more serious concern. The jury is still out on whether regularly getting into sleep debt has long-term health effects. We know that shift work and jet lag can lead to diabetes, obesity and cancer, among other problems. Catching up on sleep at the weekend, a phenomenon known as social jet lag, might cause the same kinds of health problems as shift work.

So, although anyone can recover from the short-term effects of the odd white night, a long-term habit of catching up on sleep at the weekends may well catch up with you in the end.

All of which begs the question of how to make the most of your downtime. The obvious answer is to go to bed at the right time, but even then guaranteeing good-quality sleep can be beyond our control. For instance, people often sleep poorly their first night in a new place. Studies show that this appears to be because parts of the brain remain active while people sleep. This 'night watch' effect may be an evolutionary adaptation, keeping part of your brain alert to danger.

Even in a familiar environment, sounds like a snoring partner or traffic outside can interfere with sleep, whether you're aware of them or not.

Beware blue light

Light also matters, but not necessarily in the way you might think. As well as adenosine, sleep pressure is also created by the sleep hormone melatonin. This is normally produced in the evening in response to darkness, but unfortunately LCD screens on tablets, phones and laptops generate lots of the short-wavelength blue light that suppresses melatonin production during the day. Using a screen for 2 hours in the evening reduces melatonin concentrations significantly.

That is why screen time before bed can make it harder to fall asleep. It also seems to suppress REM sleep, which is important for memory

Two sleeps

If you often wake up in the middle of the night and lie there for an hour or two before drifting back off, you may just be doing what comes naturally. According to historian Roger Ekirch of Virginia Tech, pre-industrial civilisations around the world segmented their sleep into two distinct phases, with an hour or two of 'quiet wakefulness' in the wee small hours. We've done away with this practice and call middle-of-the-night wakefulness 'insomnia'. However, studies of modern hunter-gatherers in the African and South American tropics suggest that Ekirch's idea may be wrong. Much like the rest of us, they prefer to stay up for at least 3 hours after sunset, and then sleep solidly until morning.

consolidation and emotional regulation. But if your big vice is bedtime TV, relax. The light from the box is bright, but we normally watch from far enough away to avoid the ill effects. Melatonin pills are available but probably aren't the answer. Their half-life in the body is just 30 minutes to 2 hours, which might explain why studies into whether melatonin supplements can improve sleep in general produced mixed results.

Stay cool

Temperature is another neglected factor. Melatonin cools the body by a couple of degrees while we sleep, and an overheated bedroom can interfere with this process.

And while alcohol can help you drop off, it is actually the enemy of good sleep. Having a few drinks before bed disrupts slow-wave sleep, adding a shot of alpha brainwaves that are usually only present during wakefulness. An early evening tipple can be just as bad, leading to sleep disruption in the second half of the night even after blood alcohol has returned to zero. Older people are more sensitive to the effects.

Nothing, though, is more likely to keep you awake at night than worrying about not being able to sleep. The best way to test whether you're sleep-deprived is to see how long it takes you to fall asleep at night. If you're still awake after 10 minutes, you're probably fine. If you're not, you're asleep. Which is exactly where you need to be.

12 Death

How does it feel to die?

Is it upsetting to experience consciousness slipping away, or something people can accept calmly? Are there any surprises as our existence draws to a close? None of us can know the answers for sure until our own time comes, but the few individuals who have had a brush with death interrupted can offer intriguing insights. Advances in medical science, too, have led to a better understanding of what goes on as the body gives up the ghost.

Death comes about in many ways, but it is usually a lack of oxygen to the brain that delivers the coup de grâce. It stops all electrical activity in the brain – a modern definition of death.

If the flow of oxygenated blood to the brain is stopped, people have about 10 seconds before they black out. They may take many more minutes to die, with the exact mode of death affecting the subtleties of the final experience. The details can be grisly.

Dark romance

Death by drowning, immortalised in Millais's painting of Ophelia, may have a certain dark romance about it. But in reality, it is neither pretty nor painless, though it can be surprisingly swift. Just how swift depends on several factors, such as water temperature and swimming ability. In the UK, where the water is generally cold, 55 per cent of open-water drownings occur within 3 metres of safety. Two-thirds of victims are good swimmers, suggesting that people can get into difficulties within seconds.

When victims eventually submerge, they hold their breath for as long as possible, typically 30 to 90 seconds. After that, they inhale water, splutter, cough and inhale more. Water in the lungs blocks gas exchange. Survivors have reported a tearing and burning sensation in the chest as water fills their lungs, followed by a feeling of calmness and tranquillity as they start to lose consciousness. Oxygen deprivation eventually results in the heart stopping and brain death.

Elephant in the room

Globally, heart attacks are one of the most common causes of death. The most frequent symptom is, of course, chest pain: a tightness or pressure, often described as 'an elephant on my chest'. This is the heart muscle struggling to work and possibly dying from oxygen deprivation. Pain can radiate to the jaw, throat, back, belly and arms. Other symptoms include shortness of breath, nausea and cold sweats.

The actual cause of death is often disruption of the heart's normal beat. That's because even small heart attacks can play havoc with the electrical impulses that control heart-muscle contraction. If heart-attack victims seek help, hospitals can deploy defibrillators to shock the heart back into rhythm, as well as clot-busting drugs and artery-clearing surgery. In the UK and US, more than 85 per cent of heart-attack patients who make it to hospital survive to 30 days.

The trouble is, most people who experience symptoms of a heart attack wait hours before seeking help – women more than men. And delay costs lives: once the heart stops, the 10-second countdown to unconsciousness begins. Most people who die from heart attacks never actually make it to hospital.

Exsanguination

Speed also matters if you're bleeding to death. The time it takes to die depends on where the broken blood vessels are. It can take only seconds to die if the aorta, the major vessel leading from the

Losing your head

Beheading, if somewhat gruesome, can be one of the quickest and least painful ways to die – so long as the executioner is skilled, the blade sharp and the condemned still. The pinnacle of decapitation technology is undoubtedly the guillotine. The French government adopted it in 1792 because it was seen as humane. Quick it may be, but consciousness is nevertheless believed to continue after the spinal cord is severed. A study in rats in 1991 found that it takes 2.7 seconds for the brain to consume the oxygen from the blood in the head; the equivalent figure for humans has been calculated at 7 seconds.

heart, is severed. But death can creep up slowly if the injury is to a smaller vein or artery – it can even take hours.

The average adult has about 5 litres of blood. Losses under 750 millilitres generally cause few symptoms but losing 1.5 litres causes weakness, thirst and anxiety. By 2 litres, people experience dizziness, confusion and then eventual unconsciousness. People who have survived such trauma describe a range of experiences from fear to relative calm.

Long the fate of witches and heretics, burning may be the most painful way to go. Hot smoke and flames singe eyebrows and hair and burn the throat and airways, making it hard to breathe. Burns inflict immediate and intense pain through stimulation of pain receptors in the skin. To make matters worse, they trigger a rapid inflammatory response, which boosts sensitivity to pain.

Yet most people who die in fires do not in fact die from burns but from inhaling noxious gases, such as carbon monoxide, in place of oxygen. Depending on the size of the fire and how close you are to it, carbon monoxide can start to cause headache and drowsiness in minutes, eventually leading to unconsciousness. Mercifully, many people who die in home fires are knocked out by fumes before they can wake up.

Keep your posture

A high fall is certainly among the speediest ways to die: terminal velocity (no pun intended) is about 200 kilometres per hour, achieved from a height of about 145 metres or more. According to one study, 75 per cent of victims die in the first few seconds or minutes after landing.

The natural reaction when falling is to struggle to maintain a feet-first landing, resulting in fractures to the leg bones, lower spinal column and pelvis. The impact travelling up through the body can also burst the aorta and heart chambers. Yet this is probably still the safest way to land: the feet and legs form a 'crumple zone', protecting the major internal organs.

Sky divers and other survivors of great falls often report the sensation of time slowing down. They report feeling focused, alert and driven to ensure they land in the best way possible: relaxed, legs bent and, where possible, ready to roll.

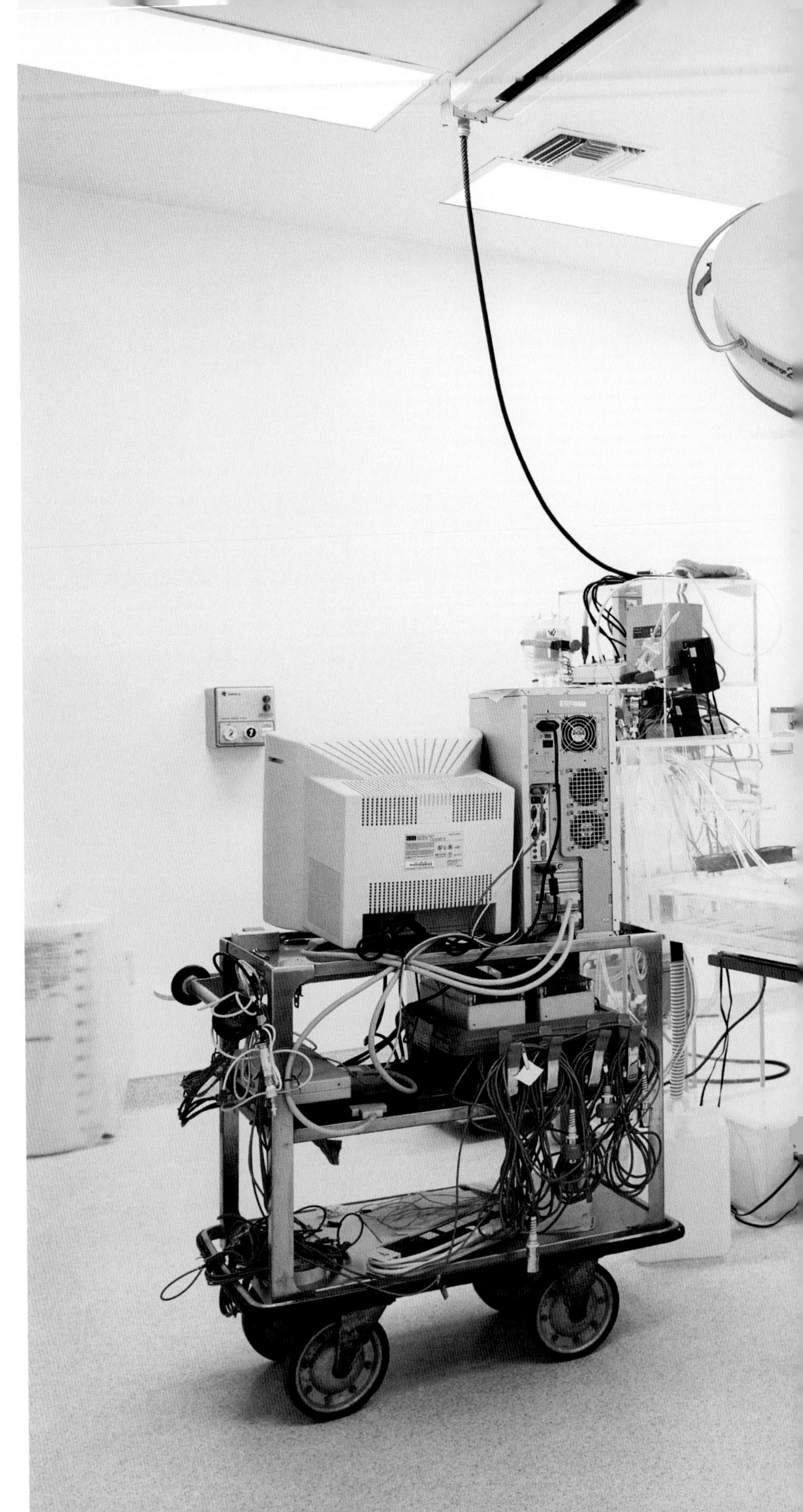

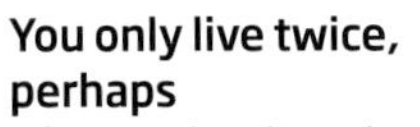

You only live twice, perhaps

After the last breath of their first life, here is where members of Alcor come to be 'de-animated'. Based in Scottsdale, Arizona, Alcor is the world's biggest cryonics company. It freezes people in the hope that future medicine will be capable of restoring them to life and curing whatever killed them first time round. As soon after death as possible, members are cooled and given drugs to prevent decomposition. Blood is replaced with medical-grade antifreeze and the bodies chilled to a frosty -196 °C. Cryonics has been described as being an afterlife for people with more money than sense. But advocates say: what if it works? Imagine the regret in the future when people look back and ask 'why did we let people die when we could have saved them?'

Credit: © Murray Ballard

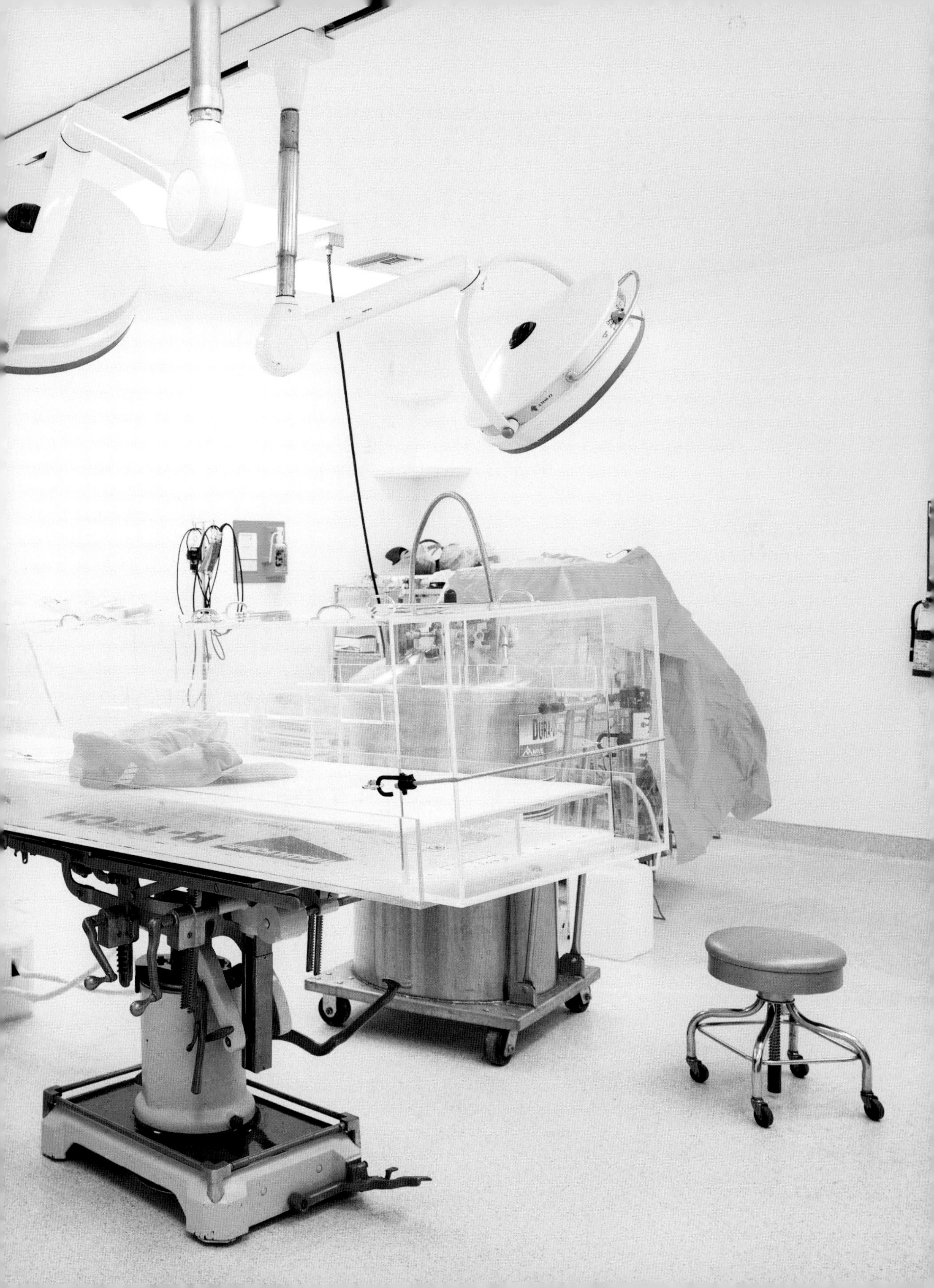

What happens to your body when you die?

It may not be nice to think about, but what happens to our mortal remains after we die is pretty fascinating. If nothing else, it proves that nature is ruthlessly efficient at clearing up its messes. Or, at least, it can be.

These days, very few of us get to be dead the old-fashioned way – out in the open, exposed to the elements. Most bodies are refrigerated soon after death or embalmed and put in a coffin. Either way, how quickly the body turns to dust depends on things like temperature, moisture and the animals, insects and microbes that happen by.

Earth to earth

In a relatively warm and moist spot with plenty of insects and scavengers, an exposed human body can be turned to bones within a few weeks and disappear completely in months. If it is given funeral rites, how quickly it decomposes will also depend on such things as how well it is embalmed, the seal on the coffin and the acidity of both the soil and the groundwater that will eventually seep in. Under these conditions, the final journey of our physical form might take anything from months to decades.

But, whatever the timescale, the vast majority of bodies will go through the same stages of decomposition.

Minutes after death

First is the 'fresh' stage. Within minutes of death, the blood acidifies as carbon dioxide builds up. As a result, cells burst open and spill enzymes which start to digest tissue from within. About half an hour later, blood starts to pool in the parts of the body closest to the ground, producing the first visible sign of decomposition. At first this looks like purplish-red blotches. Over the next day or so it turns into an almost continuous purplish mark known as livor mortis. The rest of the body turns deathly pale.

Around the same time, muscles go floppy and then stiffen as rigor mortis sets in. In life, microscopic pumps push calcium ions out of muscle cells, allowing them to relax. But in death, these stop working. Calcium ions diffuse into muscle cells from surrounding tissue, causing the muscles to contract. The body stiffens.

Rigor mortis passes after two or three days. But what looks like relaxation is actually rot setting in, as enzymes break down the proteins that held the muscles in their contracted state.

Stage two, putrefaction, gets a little ugly, and smelly. After about 48 hours, microbes get to work. These are some of the same microbes that once lived in harmony with us in our guts. Now they are fuelled by the nutrient-rich fluid that has spilled from bursting cells. The microbes churn out two compounds with names as nasty as their smells: putrescine and cadaverine give a corpse its repulsive odour.

Putrefaction appears as a green hue, spreading slowly from the belly across the chest and down the body. The colour comes from the action of bacteria that convert the blood's red haemoglobin to greenish sulphaemoglobin.

All this bacterial action also creates gases, including hydrogen, carbon dioxide, methane, ammonia, sulphur dioxide and hydrogen sulphide. These contribute to the stink and distort the body, blowing it up like a balloon and, after a month or so, bursting it open. Hydrogen sulphide also combines with iron in haemoglobin to make the black iron sulphide, which turns the skin darker.

This heralds the third stage: active decay. The rate of decomposition speeds up and what flesh is left is rapidly consumed, until all that is left is the skeleton. But something else can happen too. If the body is in particularly cold soil, 'grave wax' can form. Adipocere is a particularly spooky side effect of the work of some bacteria, such as *Clostridium perfringens*. As they digest body fat, they can leave the corpse with what looks like a wax coating.

The skeleton is the last thing to go. The hard bone minerals will not break down until they come into contact with acidic soil or water, though the process can be sped up if they are mechanically crunched by tree roots or animals. Then, once the hard parts are gone, the body's last proteins, including the collagen that once gave the bones flexibility, succumb to bacteria and fungi. They disappear.

Mummification

There are circumstances under which none of this sequence of events happens and the body doesn't decay. If it is kept completely dry, or it falls into a natural preservative like a bog, salt marsh or snow, then bacteria and enzymes can't get to work. The result is a natural mummy.

Then there are the rare cases when a person dies in the company of scavengers. In these cases, the body can be stripped to the bone and chewed into tiny pieces in days.

Ashes to ashes

Of course, without a bog, dog, shark or icy grave to hand, the only way to avoid the harsh realities of decay is cremation. At 750 °C, coffin and corpse can be burned in less than 3 hours. The ashes are then passed through a grinder to break down any big or stubborn bones.

And that, as they say, is that. However it happens, obliteration of the body is one of life's certain sequels: ashes to ashes, dust to dust, in the end there's not a lot left.

Stopping the rot

Embalming the body stops the rot in its tracks, at least temporarily. Unlike ancient Egyptian embalmers, who aimed to keep the body intact for eternity, modern embalmers keep a corpse presentable long enough for a funeral. They disinfect the body and replace the blood and other fluids with a mixture of water, dye and preservatives, usually including formaldehyde. The dye restores something like a healthy skin tone, while the formaldehyde preserves the body in several ways. It repels insects, kills bacteria and inactivates the body's enzymes. It also adds cross-links to the chains of amino acids that make up proteins, making the tissues more resistant to decomposition.

What is death?

Life is tenacious. Yet it can also be difficult to detect. Throughout history, cultures have debated how to tell for certain that somebody is dead. The results have been strange indeed . . .

3,000 YEARS AGO

ANCIENT GREECE

Amputation

The Greeks cut off a finger to ensure a person was dead before the body was cremated

250 YEARS AGO

SEVENTEENTH CENTURY

No beat, no flow

In 1628, William Harvey described how the heart pumps blood around the body, and how death arrives when heart and circulation stop

400 YEARS AGO

EIGHTEENTH CENTURY

'Back from the dead'

In 1773, William Hawes described how artificial ventilation can resuscitate apparently drowned people. Fears over death's diagnosis increased. Cardiopulmonary death came to the fore: the cessation of circulation and breathing

Bouchut's rivals put forward other schemes for confirming death, including applying leeches around the anus, pincers to the nipples, and piercing the heart with a long needle with a flag on the top

100 YEARS AGO

TWENTIETH CENTURY

Permanent cessation of all vital bodily function

Medical focus shifted from cardiopulmonary death towards the whole body. After the first kidney transplant in 1950 doctors wanted to harvest organs from dead bodies, but organs that were deprived of oxygen quickly became unusable

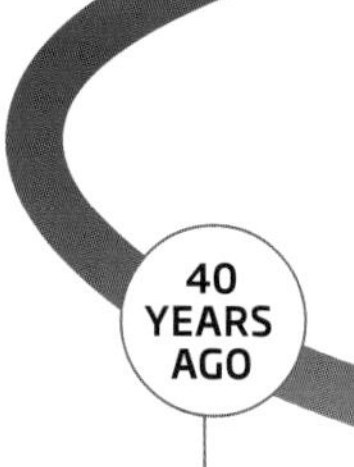

In the 1970s, studies found that the brains of 'brain-dead' people were not always dead. The Harvard tests did not confirm death of the neocortex, which produces brainwaves on an EEG. To avoid this glitch doctors were told to skip the EEG

40 YEARS AGO

1981

Brain death

The US passed an act that recognised brain death as legal death. It stated that the 'entire brain' must be dead but left examination techniques to doctors, who rarely tested the cortex

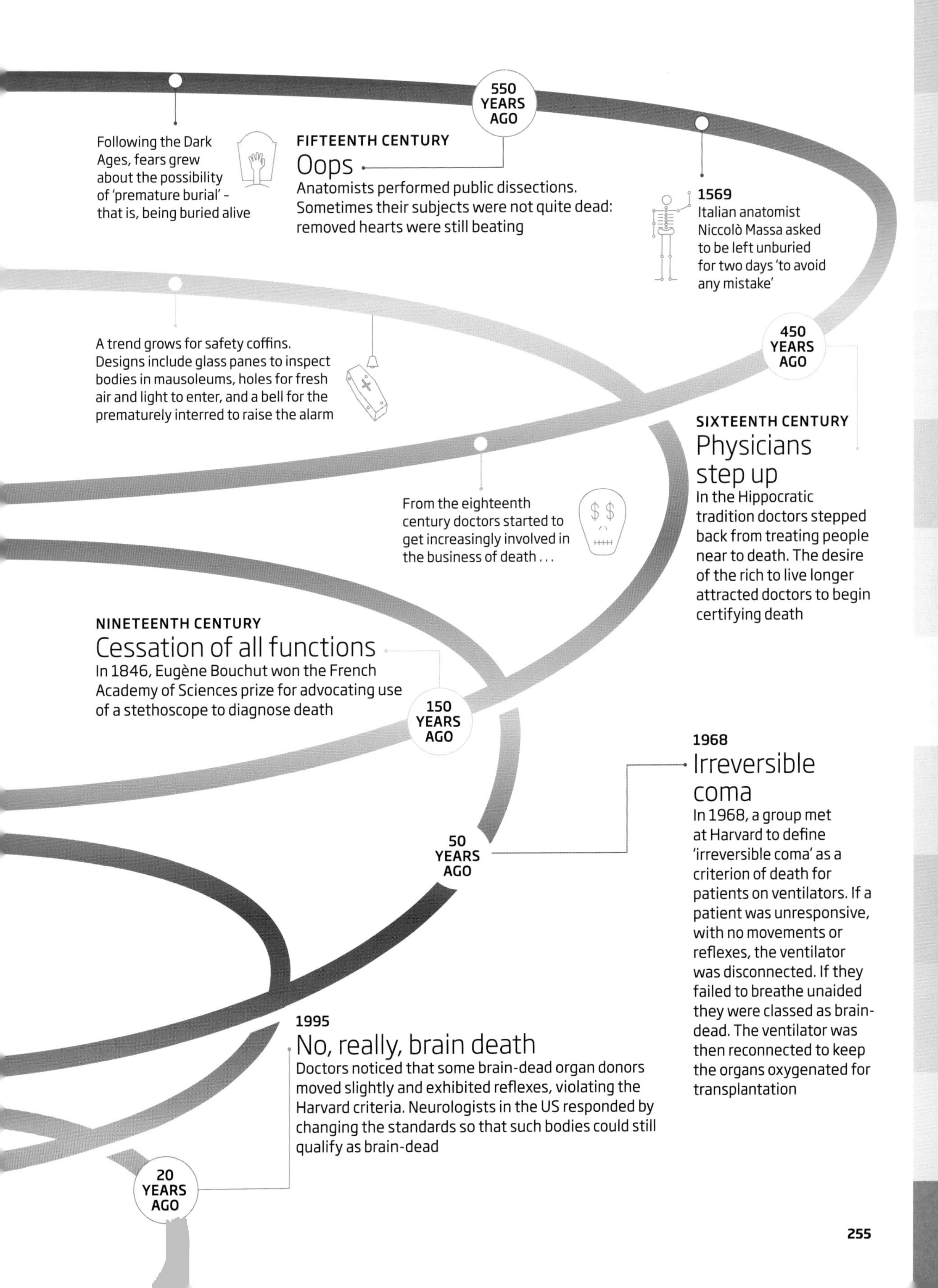

550
YEARS
AGO
Following the Dark Ages, fears grew about the possibility of 'premature burial' – that is, being buried alive
FIFTEENTH CENTURY
Oops
Anatomists performed public dissections. Sometimes their subjects were not quite dead: removed hearts were still beating
1569
Italian anatomist Niccolò Massa asked to be left unburied for two days 'to avoid any mistake'
A trend grows for safety coffins. Designs include glass panes to inspect bodies in mausoleums, holes for fresh air and light to enter, and a bell for the prematurely interred to raise the alarm
450
YEARS
AGO
SIXTEENTH CENTURY
Physicians step up
In the Hippocratic tradition doctors stepped back from treating people near to death. The desire of the rich to live longer attracted doctors to begin certifying death
From the eighteenth century doctors started to get increasingly involved in the business of death . . .
NINETEENTH CENTURY
Cessation of all functions
In 1846, Eugène Bouchut won the French Academy of Sciences prize for advocating use of a stethoscope to diagnose death
150
YEARS
AGO
1968
Irreversible coma
In 1968, a group met at Harvard to define 'irreversible coma' as a criterion of death for patients on ventilators. If a patient was unresponsive, with no movements or reflexes, the ventilator was disconnected. If they failed to breathe unaided they were classed as brain-dead. The ventilator was then reconnected to keep the organs oxygenated for transplantation
50
YEARS
AGO
1995
No, really, brain death
Doctors noticed that some brain-dead organ donors moved slightly and exhibited reflexes, violating the Harvard criteria. Neurologists in the US responded by changing the standards so that such bodies could still qualify as brain-dead
20
YEARS
AGO

The final journey of your conscious self

What happens after you die? I can name you 47 men who have tried to harness the rational horsepower of science to answer this most floaty question. Some were physicians, some physicists, some psychologists. Two were Nobel prizewinners. One is a sheep rancher. They have tackled it in labs, in hospital operating rooms, in barns behind their houses. Of them, only one, to date, has landed an irrefutable proof – not a suggestive nugget or an inexplicable anomaly, but the sort of answer you could plant your flag into and say, 'Victory! Now I know for certain.' The man's name was Thomas Lynn Bradford.

Though his background was in electrical engineering, Bradford's afterlife experiment involved gas, not electricity. On 6 February 1921, Bradford sealed the doors and windows of his rented room in Detroit, Michigan, blew out the pilot on his heater, and turned on the gas.

Two minds properly attuned

Finding out is easy. Reporting back is the challenge. For this Bradford needed an accomplice. Some weeks back, he had placed a newspaper advertisement seeking a fellow spiritualist to help him with his quest. One Ruth Doran responded. The two met and agreed, as the *New York Times* put it, 'that there was but one way to solve the mystery – two minds properly attuned, one of which must shed its earthly mantle.' The protocol was sloppy at best, for regardless of whether or not our mantle-shucking engineer came through on the telepathic wireless, Mrs Doran, for the sake of spiritualism or publicity, could simply have told the reporters that he did. But she did not lie. The *Times* ran a follow-up under the headline, 'Dead Spiritualist Silent'.

A better-pedigreed variation of the Bradford experiment was undertaken by the British physicist Oliver Lodge, once the principal of the University of Birmingham. Prior to his death in 1940, he devised the Oliver Lodge Posthumous Test. The goal, again, was to prove the existence of life after death. Lodge composed a secret message and sealed it in a packet (the Oliver Lodge Posthumous Packet) so that when, after his death, he told mediums (four of them, recruited by the Oliver Lodge Posthumous Test Committee) what the message was, their stories could be checked.

The packet itself was sealed inside seven envelopes, each envelope containing a clue the mediums could employ to jog the deceased physicist's memory should he forget his own secret. Instead, the clues merely irritated the mediums. The contents of Envelope 3, for example, read, 'If I give a number of 5 digits it may be correct, but I may say something about 2 8 0 1, and that will mean I am on the scent. It is not the real number ... but it has some connection with it. In fact it is a factor of it.' Eventually the mediums walked off the set and the Posthumous Packet was torn open, leaving the committee with nothing for their efforts but a slip of paper bearing an obscure musical fragment and a gnawing suspicion that Sir Oliver had been a few envelopes short of a stationery set.

Of course, even had the mediums succeeded, one could never be certain whether they might simply have – via some discreet Oliver Lodge Posthumous Envelope Steaming – peeked. That is why, six years later, psychologist Robert Thouless turned to the science of encryption. Thouless, the president of the venerable, endearingly daffy Society for Psychical Research and an amateur encryption expert, encoded two phrases in what he was certain was unbreakable code. He announced the project and printed the snippets in the society's journal,

Mary Roach is an author specialising in popular science and humour based in Oakland, California. She is author of *Stiff: The Curious Lives of Human Cadavers*

inviting members and mediums to try and contact him after he died and obtain from his ethereal self the key to break the code, thereby proving that one's personality survives the change of scenery known as death. Thouless died in 1984, but the phrase remains a mystery, for although some 100 people submitted what they believed to be the key, the results were invariably, to quote one Thouless Project report, 'a meaningless jumble of letters'. One party insisted that he had made contact with Thouless via no fewer than eight different mediums. Unfortunately, this man reported, Thouless had forgotten the key.

Near-death experiences

The dead-researcher approach is clearly not the way to go. A more promising tack might be to focus on those who have not quite died, but merely managed a sneak preview – in the form of a near-death experience. If someone could prove that the phenomenon is, in verifiable fact, a round-trip visit to some other dimension and not a mirage of the dying mind, that would surely be something to hang one's hopes on. But how does the person who claims to have glimpsed the beyond go about proving it? There seems to be no afterlife gift shop, no snow globes full of angel dandruff. Best to focus on one of those near-death trips that take the traveller only as far as the ceiling, enabling a reconnaissance-type view of one's corporeal hull down below. If one could at least prove that one had seen the details of the room from up there – and not remembered or hallucinated or some combination of the two – then that would at least establish the possibility of the seeming impossibility of a consciousness existing independent of its biological moorings.

And that is why there is, yes there is, a laptop computer duct-taped to the highest monitor in a cardiology operating room at the University of Virginia in Charlottesville. The computer has been programmed to show, for the duration of each operation, one of 12 images, chosen at random and unknown to anyone, including the researchers. The laptop is flat open with the screen facing the ceiling, such that the only way a surgery patient might view the image is as a disembodied consciousness. As patients come out of anaesthesia, psychologist Bruce Greyson interviews them about what they remember of their time in the operating room. So far there have been no surprises. Other, that is, than the surprising cooperation of a team of cardiac surgeons. Heart surgeons who believe that a consciousness can occasionally perceive things in an extrasensory manner, independent of a brain and eyeballs, are less rare than you might think.

But even then, how would we know that the near-death experience isn't a hallmark of dying, not death – a stopover, not a final destination? How do we know that several minutes later the bright light doesn't dim and the euphoria fade and you're just flat-out non-existent? 'We don't know,' concedes Greyson. 'It's possible it's like going to the Paris airport and thinking you've seen France.'

Life in the balance

The other way to approach the afterlife proof is to consider not the destination but the vehicle: the soul (or consciousness, if you like). If the soul were something you could weigh, like a pancreas or a wart, then proving that it abandons the corpse at death would be a simple matter of placing a dying person on a scale and watching to see if the needle went down at the moment he or she died (while also accounting for the minute amount of weight lost via moisture in exhalations and sweat).

This is exactly what a Massachusetts physician named Duncan MacDougall did, beginning in 1901, using a tricked-out industrial silk scale. His post at a tuberculosis sanatorium provided MacDougall with a steady source of study subjects. He weighed six men as they died, and there was, he said in a series of articles in *American Medicine*, always a down-tick of the needle. However, only one of his trials went off without a significant hitch. Twice the authorities barged into the room and tried to stop the proceedings. Oafish accomplices jostled the scale. Subjects died as the scale was being zeroed. And so MacDougall's claimed proof – that the soul exists, and that it weighs about 20 grams – is really no more than anecdote.

Down on the farm

Ninety-something years later, a sheep rancher in Bend, Oregon, tried to replicate MacDougall's work. When a local hospital rebuffed his solicitation for terminal patients, Lewis Hollander Jr turned to his flock. Interestingly, he found that sheep momentarily gain a small amount of weight at the moment they die. Suggesting that the answer to the question 'What happens when we die?' might in fact be: 'Our souls go into sheep.'

Of course, it's a stretch to think that the weight of a soul would register on a scale built for the likes of livestock or bolts of cloth. But what if you were to get your hands on a scale calibrated not in ounces or grams, but in picograms – trillionths of a gram? If you consider consciousness to be information energy, as some do, then it would have a (very, very, very, teeny tiny) mass. And if you were to build a closed system, such that no known sources of energy could leave or enter undetected, and you rigged it up to your picogram scale, and put a dying organism inside this system, then you could, in theory, do the MacDougall. In the course of researching a book about these various efforts to prove that there is (or isn't) an afterlife, I met a Duke University professor, Gerry Nahum, who would very much like to undertake a consciousness-weighing project of his own (offing not sheep nor men but leeches). Though he taught gynaecology and obstetrics, Nahum has a background in thermodynamics and information theory and has even worked out a 25-page proposal of exactly how to do it, if only someone will fund him the $100,000 he estimates it will cost.

One final surge

If consciousness is energy, then I suppose you don't need proof that it survives death, because proof already exists: the first law of thermodynamics – energy is neither created or destroyed. Though it's hard to take much comfort from this. Who wants to spend eternity as a blip – a gnat's fart – of disordered energy, with no brain at your disposal to help you remember or imagine or solve the Sunday crossword? What would it be like? Would there even be a *be*? Nahum uses the analogy of the computer: perhaps you'd be the operating system, stripped of its programs and interfaces. Heaven as the back of the closet where the broken-down Dells and Compaqs go.

If we are to eventually have our answer, our proof, it will no doubt come to us courtesy of quantum theory, or whatever takes its place. Few of us will understand it well enough to take much comfort, however, if indeed comfort is what it offers. I recommend that you enjoy life without worrying about the 'after' bit, and keep in mind that one day altogether too soon, bad luck or genetics will hand you the answer. In the meantime, be nice to sheep.

How well do you know yourself?

This book is full of fascinating factoids about our species. If you've already read it, these questions should be a doddle (sorry, we forgot to tell you there'd be a test . . .). If not, they should be a fun challenge that whets your appetite for the main course.

1 *Humans are unusually generous, sometimes helping total strangers for no reward. This 'true altruism' is rare in the rest of the animal kingdom, but which other animals have been found to display it?*

- ☐ **a** Vampire bats
- ☐ **b** Chimpanzees
- ☐ **c** Bees
- ☐ **d** Emperor penguins

2 *Human personality types are very varied, which is odd because natural selection tends to cut down biological variation, favouring the 'fittest'. How it is that we have such a range of personalities?*

- ☐ **a** Humans have risen above the forces of evolution
- ☐ **b** Environments change and a personality that's good for one setting may be unhelpful in others, giving natural selection a moving target
- ☐ **c** Personality is purely a product of nurture, not nature
- ☐ **d** The proportion of extroverts in society is falling as evolution gets rid of big-heads

3 *We all harbour positive illusions about ourselves. Which of the following statements is the odd one out?*

- ☐ **a** People recovering in hospital after self-inflicted car crashes often describe themselves as better-than-average drivers
- ☐ **b** Everything you do and say is being closely observed and scrutinised by those around you
- ☐ **c** Most people think they are more resistant than average to having an inflated opinion of themselves
- ☐ **d** People find a photo of themselves from a sea of faces faster if their face has been morphed to look more attractive

4 *Everybody knows that fingerprints are unique, but which other physical traits can be used to distinguish you from all the other 7 billion people on the planet?*

- ☐ **a** The exact colour of your eyes
- ☐ **b** Your face
- ☐ **c** Your heartbeat
- ☐ **d** Your bellybutton

5 *By age 45 half of men have started to go bald. What is the cause of baldness?*

- ☐ **a** Too much testosterone causes hair to fall out
- ☐ **b** Bald men are not really bald, it's just that their head hair has become wispy and invisible
- ☐ **c** There's a baldness gene that about half of men inherit
- ☐ **d** The hair follicles migrate from the top of your head to your nostrils, ears, shoulders and back

6 *A man approaches a woman in a bar and for a minute or two he notices that she touches her hair, tidies her clothes, nods and makes eye contact. What's most likely to be going on?*

- ☐ **a** The woman is signalling her interest in the man
- ☐ **b** The man is deluding himself – she is not doing any of these things
- ☐ **c** She doesn't fancy him but is playing for time to find out whether he is worth getting to know for other reasons
- ☐ **d** She is doing it for the amusement of her friend who is sat on the other side of the bar laughing

7 *Many types of cell in your body are replaced frequently. What's the newest part of you?*

- ☐ **a** The skin cells on the bottom of your feet, which wear out really fast
- ☐ **b** Your hair follicles
- ☐ **c** Your red blood cells
- ☐ **d** The cells lining your gut

8 *Brown fat has been called 'good fat' because it turns large amounts of the body's energy into heat. It is usually spurred into action by cold temperatures, but a chemical in an edible plant seems to do the same job. What is the plant?*

- ☐ **a** Celery
- ☐ **b** Durian
- ☐ **c** Chilli pepper
- ☐ **d** Frozen spinach

9 *Good executive control is seen as a good thing. It's the ability to filter out distractions, hold a train of thought and focus. Yet many people's minds are wandering about half the time. What can mind-wanderers do better than those with good executive control?*

- ☐ **a** Perform tasks that need a flash of inspiration
- ☐ **b** Relax and unwind
- ☐ **c** Learn foreign languages
- ☐ **d** Multitask

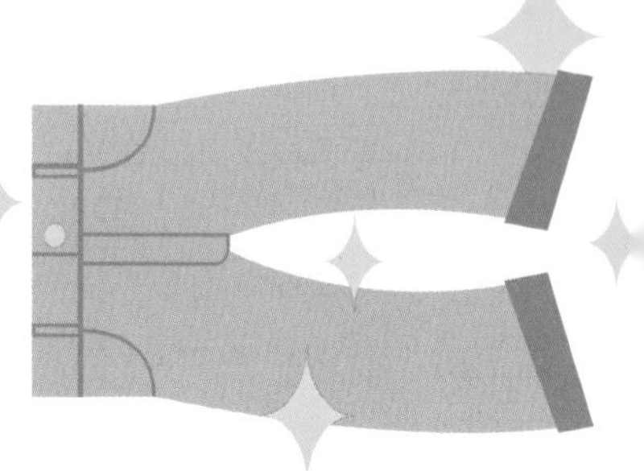

10 *The unconscious mind is by definition beyond your control, yet you could not function without it. Which of these does it not do for you?*

- ☐ **a** Make snap judgements
- ☐ **b** Act as an autopilot for day-to-day tasks such as driving and touch typing
- ☐ **c** Wake you up just before the alarm goes off in the morning
- ☐ **d** Store all your repressed urges and desires

11 *The Dunning-Kruger effect is a famous theory in psychology. What does it claim?*

- ☐ **a** That incompetent people's incompetence often causes them to fail to recognise their own incompetence
- ☐ **b** That most people are less intelligent than their parents
- ☐ **c** That emotional intelligence is more important than IQ
- ☐ **d** That psychology professors like to dream up grand theories and name them after themselves

12 *About 2 million years ago humans acquired a mutation in a gene called MYH16. How did this help to change the course of our evolution?*

- ☐ **a** Made our thick fur disappear, replacing it with thinner hair
- ☐ **b** Enabled us to tolerate alcohol
- ☐ **c** Made our tongues and lips more flexible, enabling us to speak
- ☐ **d** Made the jaw muscles weaker, which made room for the brain to expand

13 *Since we became civilised 10,000 years ago, we have continued to evolve in surprising ways. How?*

- ☐ **a** Muscles in the feet have strengthened to cope with shoes
- ☐ **b** Our fingerprints have become more swirly, for reasons nobody knows
- ☐ **c** Our sense of smell has got weaker as an adaptation for living in stinky cities
- ☐ **d** The diameters of our index fingers and nostrils have diminished at the same rate over time

14 *What is the best estimate for when humans started to wear clothes?*

- ☐ **a** 30,000 years ago, based on bone needles found in caves in France
- ☐ **b** 45,000 years ago, based on impressions of woven cloth found on the surface of clay pots in Syria
- ☐ **c** 70,000 years ago, based on the evolution of body lice, which live in clothes
- ☐ **d** 300,000 years ago, based on spun plant fibres dyed pink and orange found near Wuhan, China

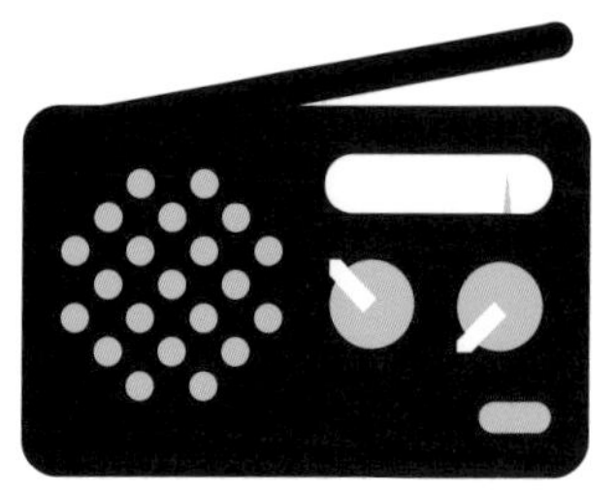

15 *Advertisers strive to exploit the pleasure we get from imagining how new possessions will improve our lives. What do psychologists call this ability?*

- ☐ **a** Anticipated material satisfaction
- ☐ **b** Transformation expectation
- ☐ **c** The saved money effect
- ☐ **d** The keeping-up-with-the-Joneses delusion

16 *Our nearest evolutionary relatives live in groups no larger than about 50, the upper limit on the number of friends an individual can groom. But we humans have expanded our group size to 150. How?*

- ☐ **a** The invention of religion made people nicer to strangers
- ☐ **b** Laughter, singing, dancing and language helped individuals to 'groom' more than one person at a time
- ☐ **c** The division of labour in hunter–gatherer societies made it essential to have more people
- ☐ **d** Inter-tribal war made it advantageous to live in larger groups

17 *Crying because we're happy or sad is a uniquely human trait. But why do we do it?*

- ☐ **a** It is a signal to others that we need help
- ☐ **b** It releases neurotransmitters that improve our mood
- ☐ **c** It rids the body of harmful chemicals
- ☐ **d** It triggers relief in the form of pain-killing opioids

18 *For some reason many people enjoy horror movies. Why?*

- ☐ **a** They help purge negative emotions that would otherwise cause a build-up of stress
- ☐ **b** They are an excuse to snuggle up to people we fancy
- ☐ **c** They are an evolutionary throwback to a time when our ancestors had to practise escaping from predators
- ☐ **d** Most of them are unintentionally funny rather than scary

19 *Why do we have no memory of events from our early childhood?*

- ☐ **a** Young children have no concept of time
- ☐ **b** You need language to remember things
- ☐ **c** Children under three years do not make the neurotransmitter dopamine, which is essential for memory
- ☐ **d** The hippocampus, which cements experiences into long-term memory, is not fully developed until age four

20 *Teenagers are notoriously hard to drag out of bed in the morning. Why?*

- ☐ **a** They can't help it – their body clocks shift and they become biological night owls
- ☐ **b** They are exhausted due to too much homework
- **c** The screens on their phones and other devices pump out blue light which makes their brains think it is daytime
- ☐ **d** They often have a hangover

Answers

1 a, 2 b, 3 b is a delusion, the others are real findings, 4 c, 5 b, 6 c (a would be correct if the woman's behaviour continued for more than four minutes), 7 d, 8 c, 9 a, 10 d, 11 a, 12 d, 13 b, 14 c, 15 b, 16 b, 17 a,18 c, 19 d, 20 a, 21 b, 22 b, 23 c, 24 a, 25 b, 26 b

21 *You can't teach an old dog new tricks. Why not?*

- ☐ **a** Older people's brains fill up with memories, making it harder to cram new ones in
- ☐ **b** It's a myth. The main problem is that adults are afraid of making fools of themselves when they try something new
- ☐ **c** The ability to master new skills peaks in childhood and diminishes rapidly as we get older
- ☐ **d** It takes thousands of hours of practice to learn new skills and adults don't have the time or patience

22 *Evolutionary theory suggests that certain techniques can maximise your chances in the dating game. Which tip has been found to work best?*

- ☐ **a** Men should go out of their way to let it be known that they are single
- ☐ **b** Wearing the colour red can increase your allure
- ☐ **c** When first chatting, men should play down their intelligence
- ☐ **d** Never try to be funny when you first meet a person you fancy

23 *Sitting down is notoriously bad for you, even for fit and healthy people. If you have to sit for long periods, how can you best minimise its harmful effects?*

- ☐ **a** Go to the gym after work
- ☐ **b** Put some tempting snacks just out of reach so you have to stretch to get them
- ☐ **c** Walk up and down for two minutes every twenty minutes
- ☐ **d** Twiddling your toes and fidgeting cancels out the effect

24 *When you've been awake a long time, which chemical builds up in your brain, increasing the desire for sleep?*

- ☐ **a** Adenosine
- ☐ **b** Dozamate
- ☐ **c** Melatonin
- ☐ **d** Napamine

25 *Power napping is a proven technique for improving alertness, concentration and attention. What should you do to maximise its effect?*

- ☐ **a** Sleep for exactly 45 minutes
- ☐ **b** Drink a cup of coffee before you drop off
- ☐ **c** Wear an eye mask and earplugs
- ☐ **d** Take micro-dose of a powerful sleeping pill

26 *Stage two of a decomposing corpse sees bacteria produce two foul-smelling chemicals. What are they called?*

- ☐ **a** Sulphanoxone and carcassine
- ☐ **b** Putrescine and cadaverine
- ☐ **c** Stenchane and rancidine
- ☐ **d** Noxigen and putrivine

Further reading

Human nature

Personality: What Makes You the Way You Are by Daniel Nettle (Oxford Landmark Science, 2009)

Born Believers: The Science of Children's Religious Belief by Justin L. Barratt (Free Press, 2012)

The Moral Landscape: How Science Can Determine Human Values by Sam Harris (Free Press, 2012)

The Crucible of Language: How Language and Mind Create Meaning by Vyvyan Evans (Cambridge University Press, 2015)

The Better Angels of Our Nature: A History of Violence and Humanity by Steven Pinker (Penguin, 2012)

The self

The Mind Club: Who Thinks, What Feels, and Why It Matters by Daniel M. Wegner and Kurt Gray (Viking, 2015)

The Man Who Wasn't There: Investigations Into the Strange New Science of the Self by Anil Ananthaswamy (Penguin, 2016)

Self Comes to Mind: Constructing the Conscious Brain by Antonio Damasio (Pantheon, 2010)

The Epigenetics Revolution: How Modern Biology Is Rewriting Our Understanding of Genetics, Disease and Inheritance by Nessa Carey (Icon Books, 2012)

Body

The Story of the Human Body: Evolution, Health and Disease by Daniel Lieberman (Allen Lane, 2013)

Surfing Uncertainty: Prediction, Action, and the Embodied Mind by Andy Clark (Oxford University Press, 2016)

The Metabolic Storm: The Science of Your Metabolism and Why It's Making You Fat (P.S. It's Not Your Fault) by Emily Cooper (Seattle Performance Medicine, 2013)

Get inside your head

Thought: A Very Short Introduction by Tim Bayne (Oxford University Press, 2013)

Thinking, Fast and Slow by Daniel Kahneman (Penguin, 2012)

The Wandering Mind: What the Brain Does When You're Not Looking by Michael Corballis (University of Chicago Press, 2015)

The Memory Illusion: Remembering, Forgetting, and the Science of False Memory by Julia Shaw (Random House, 2017)

Behave: The Biology of Humans at Our Best and Worst by Robert M Sapolsky (Bodley Head, 2017)

The Enigma of Reason: A New Theory of Human Understanding by Dan Sperber and Hugo Mercier (Penguin, 2017)

Your deep past

The Cradle of Humanity: How the Changing Landscape of Africa Made Us So Smart by Mark Maslin (Oxford University Press, 2017)

The Animal Connection: A New Perspective on What Makes Us Human by Pat Shipman (W. W. Norton, 2011)

Neanderthal Man: In Search of Lost Genomes by Svante Pääbo (Basic Books, 2014)

A Brief History of Everyone Who Ever Lived by Adam Rutherford (Weidenfeld & Nicolson, 2016)

Future Humans: Inside the Science of Our Continuing Evolution by Scott Solomon (Yale University Press, 2016)

Possessions

Spent: Sex, Evolution, and Consumer Behavior by Geoffrey Miller (Penguin, 2010)

The Marketplace of Attention: How Audiences Take Shape in a Digital Age by James Webster (MIT Press, 2014)

Stuff by Daniel Miller (Polity Press, 2009)

Evocative Objects: Things We Think With by Sherry Turkle (MIT Press, 2011)

Friends and relations

How Many Friends Does One Person Need?: Dunbar's Number and Other Evolutionary Quirks by Robin Dunbar (Faber & Faber, 2011)

Thinking Big: How the Evolution of Social Life Shaped the Human Mind by Robin Dunbar, Clive Gamble and John Gowlett (Thames and Hudson, 2014)

Touch: The Science of Hand, Heart and Mind by David J. Linden (Viking, 2015)

Emotions

How Emotions Are Made: The Secret Life of the Brain by Lisa Feldman Barrett (Houghton Mifflin Harcourt, 2017)

Emotion: Pleasure and Pain in the Brain by Morten Kringelbach and Helen Phillips (Oxford University Press, 2014)

Don't Look, Don't Touch, Don't Eat: The Science Behind Revulsion by Valerie Curtis (University of Chicago Press, 2013)

The Nostalgia Factory: Memory, Time and Ageing by Douwe Draaisma (Yale University Press, 2013)

Life stages

The Gardener and the Carpenter: What the New Science of Child Development Tells Us About the Relationship Between Parents and Children by Alison Gopnik (Farrar, Straus and Giroux, 2106)

Teenagers: A Natural History by David Bainbridge (Portobello Books, 2010)

Middle Age: A Natural History, by David Bainbridge (Portobello Books, 2013)

Sex and gender

The Essential Difference: Men, Women and the Extreme Male Brain by Simon Baron-Cohen (Penguin, 2012)

Testosterone Rex: Unmaking the Myths of Our Gendered Minds by Cordelia Fine (Icon Books, 2017)

Brain Storm: The Flaws in the Science of Sex Differences by Rebecca M. Jordan-Young (Harvard University Press, 2010)

Well-being

Cure: A Journey Into the Science of Mind Over Body by Jo Marchant (Canongate, 2016)

Why We Get Fat: And What to Do About It by Gary Taubes (Vintage, 2012)

How Not to Die: Discover the Foods Scientifically Proven to Prevent and Reverse Disease by Michael Greger (Flatiron, 2015)

Obesity: The Biography by Sander L. Gilman (Oxford University Press, 2010)

Death

Stiff: The Curious Lives of Human Cadavers by Mary Roach (Penguin, 2004)

Spook: Science Tackles the Afterlife by Mary Roach (W. W. Norton, 2006)

Smoke Gets in Your Eyes: And Other Lessons from the Crematorium by Caitlin Doughty (Canongate, 2016)

The Worm at the Core: On the Role of Death in Life by Sheldon Solomon, Jeff Greenberg and Tom Pyszczynski (Random House, 2015)

Guest authors

Daniel Nettle is professor of behavioural science at Newcastle University, UK. He studies topics relating to behaviour, ageing and health in humans and in non-human animals. He is author of *Personality: What Makes You the Way You Are*

Justin L. Barrett is professor of psychology at Fuller Theological Seminary in Pasadena, California. He is a founder of the new field looking at the cognitive science of religion and author of several books, including *Born Believers: The Science of Children's Religious Belief*

Jan Westerhoff is associate professor of religious ethics at the University of Oxford, UK. His research focuses on contemporary metaphysics and on Indian Buddhist philosophy. He is author of several books, including *Reality: A Very Short Introduction*

Tony Prescott is professor of cognitive neuroscience at the University of Sheffield, UK. Within his research he focuses on bio-inspired robotics with a long-term focus on understanding mammalian brain architecture

Tim Bayne is professor of philosophy at Monash University in Melbourne, Australia. He concentrates on the philosophy of mind and cognitive science, with a particular interest in the nature of consciousness. He is author of *Thought: A Very Short Introduction*

Pat Shipman is retired adjunct professor of anthropology at Pennsylvania State University, University Park. She is author of numerous articles for *New Scientist* and several books including *The Animal Connection: A New Perspective on What Makes Us Human*

Geoffrey Miller is associate professor of psychology at the University of New Mexico, Albuquerque. An evolutionary psychologist, he is known for his expertise in sexual selection in human evolution and is author of *Spent: Sex, Evolution, and Consumer Behavior*

Robin Dunbar is professor of evolutionary psychology at the University of Oxford, UK. His research focuses on the evolution of sociality in primates and other mammals, especially group bonding and size. He is author of *How Many Friends Does One Person Need?: Dunbar's Number and Other Evolutionary Quirks*

Lauren Brent is lecturer in animal behaviour at the University of Exeter, UK. Her research asks whether interpersonal relationships impact upon the longevity and reproductive success of group-living mammals.

David Bainbridge is clinical veterinary anatomist at the University of Cambridge, UK. He is active in promoting the wider understanding of science and has written several books, such as *Middle Age: A Natural History* and *Teenagers: A Natural History*

Jo Marchant is a London-based science journalist and author. She has worked as a writer and editor at *New Scientist* and *Nature*. Her most recent book is *CURE: A Journey Into the Science of Mind Over Body*

Mary Roach is an author specialising in popular science and humour based in Oakland, California. Among her books are *Stiff : The Curious Lives of Human Cadavers* and *Spook: Science Tackles the Afterlife*

Acknowledgements

This book would not have happened without the support, ideas and hard work of a lot of people at *New Scientist*. Thanks to Kate Douglas, who helped create the concept and flesh out the contents; to Catherine Brahic, Tiffany O'Callaghan and Caroline Williams, who supplied vital editing support at short notice; to editor-in-chief Sumit Paul-Choudhury and publisher John MacFarlane for their unstinting support; to our agent Toby Mundy for sage advice; and to everyone else at *New Scientist* for their constant curiosity and journalistic brilliance.

The team at John Murray also deserve huge thanks, especially Georgina Laycock and Kate Craigie for their ideas, energy and persistence. Thanks also to Nick Davies, Will Speed for the cover, Mandi Jones in production, Nicky Barneby for design, Yassine Belkacemi and Jess Kim for publicity and marketing, Anna Alexander for rights, and Ben Gutcher, Lucy Hale and Sarah Clay in sales.

On the other side of the Atlantic, we are hugely indebted to Jennifer Daniel who did a heroic and fabulous job of turning scientific ideas into accessible and entertaining graphics. Thanks also to Robyn Kanner of Transhealth for her guidance and expertise on the infographic 'A world of gender identity'.

Some of the material in this book is adapted from articles previously published in *New Scientist*.

Every reasonable effort has been made to trace the copyright holders, but if there are any errors or omission, John Murray will be pleased to insert the appropriate acknowledgement in any subsequent printings or editions.

Index

First published in Great Britain in 2017 by John Murray (Publishers)
An Hachette UK company

First published in USA in 2017 by Nicholas Brealey Publishing

1

A CIP catalogue record for this title is available from the British Library

John Murray, UK
ISBN 978 1 473 62928 8
Ebook ISBN 978 1 473 62927 1

Nicholas Brealey Publishing, USA
ISBN 978 1 473 65870 7
Ebook ISBN 978 1 473 65871 4

Design: Nicky Barneby
Picture Editor: Kirstin Kidd
Additional writing by Kate Douglas, Caroline Williams and Catherine Brahic
Copy editor: Ned Pennant-Rea
Proof readers: Steve Cox and Chris Simms
Indexer: Ruth Ellis

Typeset in Monotype TheAntiqua and Monotype Soho Gothic by Barneby Ltd

Printed and bound in Germany by Mohn Media GmbH

John Murray policy is to use papers that are natural, renewable and recyclable products and made from wood grown in sustainable forests. The logging and manufacturing processes are expected to conform to the environmental regulations of the country of origin.

John Murray (Publishers)
Carmelite House
50 Victoria Embankment
London EC4Y 0DZ

www.johnmurray.co.uk

Nicholas Brealey Publishing
Hachette Book Group
Market Place Center, 53 State Street,
Boston, MA 02109, USA

www.nicholasbrealey.com